how to
know the
spring
flowers

The **Pictured Key Nature Series** has been published since 1944 by the Wm. C. Brown Company. The series was initiated in 1937 by the late Dr. H. E. Jaques, Professor Emeritus of Biology at Iowa Wesleyan University. Dr. Jaques' dedication to the interest of nature lovers in every walk of life has resulted in the prominent place this series fills for all who wonder **"How to Know."**

John F. Bamrick and Edward T. Cawley
Consulting Editors

The Pictured Key Nature Series

How to Know the

AQUATIC INSECTS, Lehmkuhl
AQUATIC PLANTS, Prescott, Second Edition
BEETLES, Arnett-Downie-Jaques, Second Edition
BUTTERFLIES, Ehrlich
FALL FLOWERS, Cuthbert
FERNS AND FERN ALLIES, Mickel
FRESHWATER ALGAE, Prescott, Third Edition
FRESHWATER FISHES, Eddy-Underhill, Third Edition
GILLED MUSHROOMS, Smith-Smith-Weber
GRASSES, Pohl, Third Edition
IMMATURE INSECTS, Chu
INSECTS, Bland-Jaques, Third Edition
LICHENS, Hale, Second Edition
LIVING THINGS, Winchester-Jaques, Second Edition
MAMMALS, Booth, Fourth Edition
MITES AND TICKS, McDaniel
MOSSES AND LIVERWORTS, Conard-Redfearn, Second Edition
NON-GILLED MUSHROOMS, Smith-Smith-Weber, Second Edition
PLANT FAMILIES, Jaques, Second Edition
POLLEN AND SPORES, Kapp
PROTOZOA, Jahn, Bovee, Jahn, Third Edition
SEAWEEDS, Abbott-Dawson, Second Edition
SEED PLANTS, Cronquist

SPIDERS, Kaston, Third Edition
SPRING FLOWERS, Verhoek-Cuthbert, Second Edition
TREES, Miller-Jaques, Third Edition
TRUE BUGS, Slater-Baranowski
TRUE SLIME MOLDS, Farr
WEEDS, Wilkinson-Jaques, Third Edition
WESTERN TREES, Baerg, Second Edition

how to know the
spring flowers

Second Edition

Susan Verhoek
Lebanon Valley College

Mabel Jaques Cuthbert

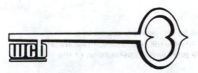

The Pictured Key Nature Series
Wm. C. Brown Company Publishers
Dubuque, Iowa

Line drawings in figures 151, 152, 153 are reprinted from NATIVE ORCHIDS OF NORTH AMERICA: NORTH OF MEXICO by Donovan S. Correll, with permission of the publishers, Stanford University Press. © 1951, 1978 by the Board of Trustees of the Leland Stanford Junior University.

Line drawings in figures 109, 135, 186, 230, 249, 338 are from MANUAL OF THE VASCULAR FLORA OF THE CAROLINAS, by Albert E. Radford, Harry E. Ahles, and C. Ritchie Bell. Copyright 1968 The University of North Carolina Press. Reprinted by permission of the publisher.

Line drawings in figures 319, 322, 327 are from *Violets (Viola) of Central and Eastern United States: an Introductory Survey* by Norman H. Russell, SIDA 2 (1). 1965. Used by permission.

Copyright 1970, 1982 by The McGraw-Hill Companies, Inc.
Copyright 1949 by N. E. Jaques

Library of Congress Catalog Number: 81-68818

ISBN 0-697-04782-2

Printed in the United States of America
10 9 8 7

To my parents,

who began and encouraged my interest in spring flowers.

Contents

Preface ix

Introduction 1
  Plant Biology 1
  Plant Groups 6
  Botanical Names 7
  Edible, Medicinal and Poisonous Plants 8
  Collecting and Preserving 10
  Endangered Species 14
  Measurements 16
How to Use the Keys 17
Keys to Major Groups and Families 18
  Monocotyledons, Genus and Species Keys 30
  Dicotyledons, Genus and Species Keys 71

List of Families in Phylogenetic Order 226
List of Families by Common Name 227
Alphabetic Index of Genera by Family 229
Index and Pictured Glossary 231

Preface

This edition of *How to Know the Spring Flowers*, like the first edition, is intended to allow the easy identification of commonly encountered spring flowering herbaceous plants in North America east of the Rocky Mountains (exclusive of tropical Florida). The plants chosen for inclusion in this edition begin to bloom between February and the first week in June. The species treated here were selected because of their frequency and widespread occurrence, based on a tabulation of species in various state and regional spring floras.

The text and introductory material have been largely rewritten. Drawings of 74 new species have been added and 53 others mentioned, among these some frequently encountered weeds, and a few rare species have been omitted, increasing the total number of species and varieties to 505 in this edition.

The organization of this edition has been changed to emphasize plant families. An initial key identifies plants by family in a semi-diagnostic key. Following that, each family has a separate key to species. The arrangement of family treatments follows basically that of the manuals listed in the following paragraph. The main key characters used are those of leaves and flowers.

Descriptions of species are based on original observations of plants either in the field or as herbarium specimens, and on a composite of information from Henry A. Gleason and Arthur Cronquist, *Manual of Vascular Plants of Northeastern United States and Adjacent Canada,* 1963; Albert E. Radford, Harry E. Ahles, and C. Ritchie Bell, *Manual of the Vascular Flora of the Carolinas,* 1968; Wilbur H. Duncan and Leonard Foote, *Wildflowers of the Southeastern United States,* 1975; Janét Bare, *Wildflowers and Weeds of Kansas,* 1979; and from state floras or spring flower manuals. The distribution maps have been revised and are as accurate as data from the manuals and floras allow. The nomenclature has been brought up to date and follows John T. Kartesz and Rosemarie Kartesz, *A Synonymized Checklist of the Vascular Flora of the United States, Canada, and Greenland,* 1980.

New types of information have been added. The Introduction includes a brief section on endangered species and a prologue to the medical, edible or poisonous properties cited in the plant descriptions. An expanded section on plant biology lays the groundwork for understanding the significance of the reports of insect visitors and pollinators to the flowers.

This book maintains the pictured key approach and is midway in difficulty between wildflower books that group flowers simply by flower color and the more technical state and regional manuals. It allows easy identification of a plant but also encourages recognition of plant families and familiarity with keying techniques. Its wide geographical scope permits use both at home and on field trips or vacations. As such, the book is useful to botanists, ecologists and serious students of nature.

At the same time, the elementary introductory material provides an understanding of basic structural botany and related aspects of plant identification, so that it is suitable for use by beginning students of botany and neophyte naturalists. The accounts of useful or poisonous properties of the plants add to a well-rounded appreciation of the spring flowers by professionals and hobbyists alike.

By means of their expertise and good will, a group of people and institutions have helped this edition toward completion. The L. H. Bailey Hortorium, Cornell University, made its collections and office space available to me during two summers, and made me feel welcome. Wayne Perry painstakingly prepared most of the new line drawings for this edition. Joan Bernardo carried out the difficult task of collating and typing two parallel drafts into the final manuscript. Diane Miske checked the previous distribution maps and gathered data from which the new and revised maps were prepared. These four deserve many heartfelt thanks. The book could not have been put into final form without additional illustrations from several sources. Four new illustrations in the monocots were drawn by R. Scott Bennett. Drawings for figures 109, 135, 186, 230, 249, and 338, are from Radford, A. E., H. E. Ahles, and C. R. Bell, *Manual of the Vascular Flora of the Carolinas*, copyright 1968 by The University of North Carolina Press; figures 151, 152, and 153, are from Correll, D. S., *Native Orchids of North America North of Mexico*, 1978, Stanford University Press; figures 319, 322, and 327, are from Russell, N. H., *Violets (Viola) of Central and Eastern United States: an Introductory Survey*, 1965, SIDA, Volume 2, Number 1. All of these are used by permission of the copyright holder. A series of drawings were also taken from Stevens, G. T., *Illustrated Guide to Flowering Plants*, 1910, for which reprint permissions were requested but the copyright holder could not be located.

Bruce MacBryde, U. S. Fish and Wildlife Service, kindly reviewed the section on endangered species and a number of other people read and offered comments on the manuscript. To them thanks are also due.

I joyfully remember the thrill of finding, and identifying, a Pasque Flower after a long Minnesota winter. This book is the second edition of the book by Mabel Cuthbert I had in hand that warm spring day. I hope that it continues the tradition of excitement of first-time discoveries and the continuing satisfaction of knowing the answer to the annual question "what kind of flower is that?"

Susan Verhoek
Annville, Pennsylvania, January 15, 1981

Introduction

Spring flowers are eagerly watched for as proof that a new growing season has come. Since each flower is an important symbol of springtime and the blooming of each new species is an event in the progress toward summer, most people observe the spring flowers more closely than the flowers at any other season. They notice differences in the size, shape, color, and number of parts of the flowers.

Once these characteristics have been observed, they can be used to discover the name of the plant. Just as appreciation of a painting, a piece of music or a ball game is better if a person knows something about the subject, enjoyment of the flowers in springtime can be greater if a flower is known by its special name and if its family relationship to other plants is realized.

As a background to identifying the plants by name, some general information about flowering plants is helpful.

PLANT BIOLOGY

The characteristics by which plants are identified can come from any part of a plant. The plant has four main organs: root, stem, leaf, and flower. The *root* absorbs water and minerals from the soil, and anchors the plant in the ground. The *stem* is the structure which supports the leaves and flowers, and contains the water and food transport channels between the leaves and flowers and the root. The *leaves* are the food producing organs of the plant which take energy from the sun and use it to convert, through photosynthesis, simpler chemicals in the plant into carbohydrates. The *flower* is the reproductive organ. Its main function is to produce seed for the next generation. A flower may be a complex structure, made up of many parts, and may be very different from one species to the next. For identification of common spring flowers, leaf and flower characters are the most easily seen. They are the main characters used in the following keys.

The Leaf

Leaves are characterized by their overall shape, by the shape of the tips and leaf bases, and by their margins (Fig. 1).

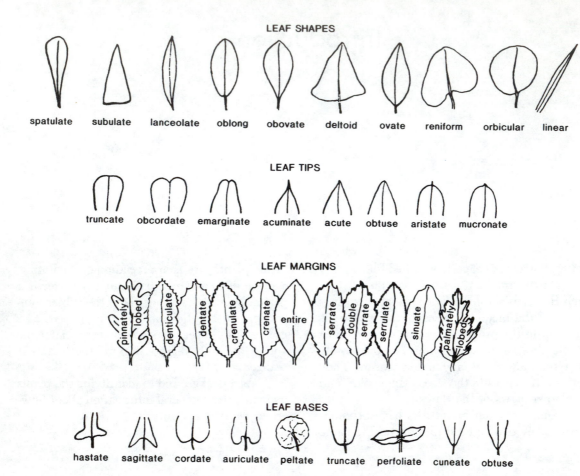

LEAF SHAPES

spatulate · subulate · lanceolate · oblong · obovate · deltoid · ovate · reniform · orbicular · linear

LEAF TIPS

truncate · obcordate · emarginate · acuminate · acute · obtuse · aristate · mucronate

LEAF MARGINS

pinnately lobed · denticulate · dentate · crenulate · crenate · entire · serrate · double serrate · serrulate · sinuate · palmately lobed

LEAF BASES

hastate · sagittate · cordate · auriculate · peltate · truncate · perfoliate · cuneate · obtuse

Figure 1

Simple leaves consist of a single leaf-like portion, the *blade,* attached to the stem by a stalk, the *petiole* (Fig. 2). Occasionally there are two small leaf-like structures, the *stipules,* at the base of the petiole.

A *compound* leaf is one in which the blade of the leaf is so deeply lobed that in fact the leaf is made up of individual leaf-like structures called *leaflets.* The main stalk of the leaf on which the leaflets are borne is called the *rachis.* Compound leaves may be pinnately or palmately compound, depending upon the arrangement of the leaflets. A palmately com-

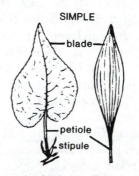

SIMPLE

blade — petiole — stipule

net-veined   parallel-veined

Figure 2

pound leaf has the leaflets all attached to the rachis at the same point (Fig. 3). In a pinnately compound leaf the leaflets are attached on opposite sides along the rachis so that the leaf looks like a feather (Fig. 3). To find the clue to whether the leaf is simple or compound look for a small bud emerging from the stem just at the base of the petiole. A true leaf has the bud at its base. If the blade belongs to a leaflet, there will be no bud at its base, but at the base of the rachis.

COMPOUND

palmate

pinnate

Figure 3

In the blade of the leaf are thickened strands, the *veins,* which consist of the food and water conducting tissues. The pattern of the veins may be palmate, pinnate from the main central vein (the midrib), parallel, or net-like. The terms pinnate and palmate venation describe the same pattern as the leaflet arrangement in compound leaves.

The point at which the leaf is attached to the stem is called the *node.* The portion of the stem between the nodes is the *internode.* The arrangement of the leaves on the stem may be opposite, with a pair of leaves present at any one point on a stem; whorled, with three or more leaves present at a node; or alternate, with only one leaf present at a node (Fig. 4).

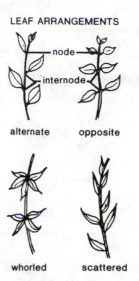

LEAF ARRANGEMENTS

node

internode

alternate   opposite

whorled   scattered

Figure 4

The Flower

A complete flower (Fig. 5) has four sets of parts. From the outside to the inside they are: 1. a *calyx* composed of *sepals,* 2. a *corolla* composed of *petals,* 3. *stamens* that may be separate or united, 4. a *pistil* of one to many carpels. These parts are borne on the broadened tip (receptacle) of the flower stalk (peduncle).

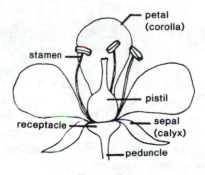

petal
(corolla)

stamen

pistil

receptacle

sepal
(calyx)

peduncle

Figure 5

In some flowers the calyx is united into a tube or the petals make a tubular or bell-shaped corolla (Fig. 6). When only one set of floral envelopes (sepals and petals) is present, it is arbitrarily called the calyx. The calyx and the corolla together are known as the *perianth*.

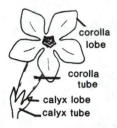

Figure 6

A perfect flower has both stamens and pistils with or without sepals and petals. These are the only parts that actually function to produce seed. A stamen consists of a pollen-bearing *anther* which is usually supported by a *filament* (Fig. 7a).

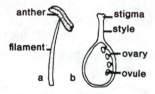

Figure 7

A pistil (Fig. 7b) must have an *ovary* and *stigma*. A stem-like *style* often separates these two parts. The basal ovary contains the *ovules* which ripen into seeds after fertilization. The stigma is the tip that receives the pollen; it is connected to the ovary by the style, which varies in length and may even be absent. A *carpel* is a simple pistil or one section of a compound pistil (Fig. 8).

The stamens are the male part of the plant, and the pistil is the female part. A flower may have stamens only (staminate) or pistils

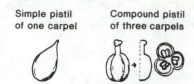

Figure 8

only (pistillate) (Fig. 9). These unisexual flowers are termed imperfect. In a dioecious species two separate plants bear staminate and pistillate flowers. A monoecious species has both staminate and pistillate flowers on the same plant.

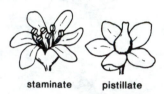

Figure 9

Flowers are either solitary or clustered. The cluster of flowers is called the *inflorescence*. Figure 10 shows the most common inflorescence types.

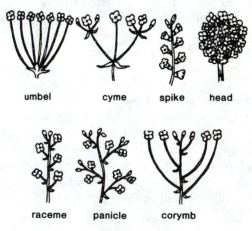

Figure 10

The Fruit and Seeds

The mature ovary with its contents the seeds, is the *fruit*. Fruits may be dry or fleshy and may contain one to many seeds. In a few species of plants the ovary wall will mature into a seedless fruit if the ovules are not fertilized (as in seedless oranges). Ordinarily, however, the entire ovary dies if one or more seeds do not develop within it.

Dehiscence is the method by which a dry fruit opens to discharge its seeds. Among dehiscent fruits are the violet capsule, mustard silique, bean and pea pod (also called legume), larkspur and milkweed follicle. Some dry fruits are indehiscent, or do not open to discharge the seed. Examples of indehiscent fruits are: the achenes of buttercup, anemone, and dandelion, and the nutlets of mint and borage (Fig. 11).

DEHISCENT

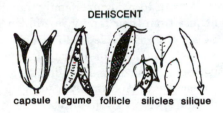

capsule   legume   follicle   silicles   silique

INDEHISCENT

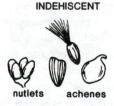

nutlets   achenes

Figure 11

Fleshy fruits include the drupe of dogwood, the berry of tomato, Solomon's seal, and asparagus. The strawberry is an enlarged fleshy receptacle with achenes imbedded in the surface. This type of fruit is called an accessory fruit.

The seed is a partially-developed young plant with embryonic root, stem and leaves plus food reserves. This embryo plant is well-protected by the hard outer seed-coat and has an ample food supply to keep it going until it develops green leaves of its own and can manufacture food for itself. Seed dispersal is accomplished by wind, water, and animals.

The Function of the Flower in Reproduction

Just as in animals, reproduction occurs in plants when a sperm nucleus unites with the egg cell. The sperm nucleus is carried in the pollen grain. The egg cell is contained in the ovule, which is the immature seed. The ovule is in turn contained and protected by the lower portion of the pistil (the ovary) at the center of the flower.

The pollen grain must be carried from the male part of the flower (the anther) to the top of the pistil before reproduction can take place. In some flowers the pollen is carried by wind or raindrops. Among other species it is necessary to have an insect, hummingbird, or even a bat, carry pollen from flower to flower. The hummingbirds, bats, and insects visit the flowers in search of pollen to eat or nectar to drink. The flowers that depend on birds and insects to carry their pollen have odors or showy petals which advertise the food and drink available within them. The color and pattern of flowers are recognized by the pollinators and are just as effective as advertising symbols to insects and birds as are the distinctive signs of fast food restaurants to human beings.

When the insects or birds visit a flower to feed, their bodies are dusted with pollen and they also brush the pollen onto the stigmatic tip of the pistil of the flower. Once the pollen is on the stigma of a flower of the same species, the pollen grain germinates and a tube grows down through the pistil until it reaches the egg cell in the ovule. The egg cell is then fertilized, giving rise to the embryo for the next generation of plants, and the seed which contains the young embryo begins to develop. As the seed develops it causes the ovary of the pistil to develop into the fruit. The fruit is the matured female part of the flower.

PLANT GROUPS

The plants identified in this book are a particular portion of the plant kingdom. These plants all produce flowers in the spring and they are all herbaceous. A herbaceous plant is one without any woody parts. After the growing season these plants either put all of their energy into fruits and seeds which lie dormant through the winter after the whole plant dies, or else the above-ground parts die back to the ground and the nutrients are stored in the root.

Plants form one of the major groups of living things. They are a group on a level with the other kingdoms which contain animals, fungi, and single-celled organisms like bacteria and protozoa. The organisms called plants are divided into subgroups, some for the seaweeds and other algae, and others for the mosses and liverworts, for the ferns, and for the seed-bearing plants. The seed plants are divided further into classes for plants which produce seeds in cones instead of in flowers, such as pine trees, and into a class for plants with flowers. The plants with flowers are called Angiosperms. It is the herbaceous angiosperm flowers that bloom in the spring that this book identifies.

The group for Angiosperms is further divided into smaller and smaller groups, representing even finer distinctions in classification. With each smaller group, genetic relationships of the members within the groups become ever closer. There is something of a parallel in human family structure. In humans there are extended families of the type in which 200 people return to a home town for the annual family reunion. At these reunions it may be possible to trace a particular family trait. Within each large family are the smaller family units in which cousins can be identi-

fied because they all look somewhat alike. And finally, individual family groups can be sorted out because some of the cousins look more alike than other sets of cousins do. Among Angiosperms at the "superfamily" level are two subclasses, the Monocotyledons, for flowers with parts in threes and one food storage structure in the seed, and the Dicotyledons, for flowers with parts in fours or fives and two food storage structures in the seed. These two subclasses are divided into orders, and the orders into families. Members of families of plants are closely enough related so that it is possible to learn to recognize family members by sight just from their family resemblance.

Each family is divided into *genera* (singular, *genus*). Members of a genus look more like each other than they look like members of any other genus. And finally, each genus is divided into *species*. Each different kind of plant that we recognize is a different species. For example, we recognize a certain rose as species "setigera." The genus name of roses is the Latin form *Rosa*. To indicate that "setigera" is a rose and to recognize the relationship of species "setigera" to other species of roses we refer to it by its double name *Rosa setigera*. The genus *Rosa* is in the family Rosaceae. Other genera related to *Rosa* are also included in the family Rosaceae. Some of these are *Rubus* (the blackberry), and *Pyrus* (the pear).

Occasionally a species is further subdivided into varieties. The abbreviation *var.* after a species name indicates that the next name is the variety name and that within the species there are slight genetic variations in the characteristics of the species which it is useful for us to identify. Some examples of varieties are those in which leaves of a normally pointed-leaved species have rounded tips, or a species usually characterized by thorns has a number of thornless members.

The hierarchy of plant classification is as follows, with the Prairie Rose, *Rosa setigera* Michx. var. *tomentosa* Torr. & A. Gray, as an example:

Kingdom—Plantae
Class—Angiospermae
Subclass—Dicotyledonae
Order—Rosales
Family—Rosaceae
Genus—*Rosa*
Species—*setigera*
Variety—var. *tomentosa*

In the biological world there is also a certain amount of random variation, some of which is brought about by differences in environment. In identifying plants it is well to keep this in mind, and look for a typical specimen rather than the largest one in the population. For this reason, measurements in the keys are often given as a range of sizes that plants in a species may attain.

BOTANICAL NAMES

The scientific name by which a plant is known is made up of three parts. The first two are the actual name of the plant species and are Latin in form. The first word of the name is the *genus*. It is written with a capital letter. The second word is the specific epithet, the name of the *species*. It is written with a small letter. Scientific names are underlined or printed in italic type.

The use of the two names together is called binomial nomenclature. This was an improvement in nomenclature popularized in the 1750's by the Swedish botanist Linnaeus. Before that time each plant was named by a whole phrase which described the plant in Latin.

The descriptive value of names has not been lost even though the whole phrase is no longer used. Many species names still describe characteristics of the plant. For example, the buttercup species *Ranunculus bulbosus* has a bulbous base; *Myosurus minimus* is a tiny plant. Species names may describe the locality where the plant was found by such names as "californicus, canadense, pennsylvanica, and carolinianum." Names of species often describe aspects of the plant itself: "hirsuta" means a hairy plant, "caerulea" refers to a plant with sky-blue flowers and "diphylla" is a species with two leaves. The name "officinalis" indicates that at one time the plant was on the official list of plants used by pharmacists. "Edulis" means that the plant was a food plant and "sativus" means that it was cultivated as a crop. Occasionally, although not often with the common spring flowers, a species is named in honor of a botanist, as for example *Zigadenus nuttallii,* named in honor of Thomas Nuttall, who made many observations on American plants in the early 1800's. When a species is named after a person, some botanists prefer to capitalize the species name also.

The third part of the plant name is an abbreviation which indicates the last name of the person or persons who gave the species its binomial. This is the "author name." The author name is included because it helps to avoid confusion between similar species names and because it adds to the history and background of the species to know which specialist provided the name for it. One common abbreviation found after many plants of the Northeast is "L." This is the abbreviation for Linnaeus, who did most of his work in botany just as the eastern United States and Canada were being explored. Michaux (Michx.) and

Muhlenberg (Muhl.) were later botanists who wrote about American plants.

Often after a species binomial there will be two name abbreviations, one of them in parenthesis. This means that there has been some shifting in the classification of the species as a result of further research. The species was named originally by the author whose name is in parenthesis but later the species was shifted from the original genus to another one. The author who made the shift is indicated by the name which follows the parenthesis. For example, *Draba reptans* (Lam.) Fern. indicates that the species *reptans* was transferred to the genus *Draba* by the botanist Fernald.

Each plant has only one scientific name and it is the same in all parts of the world and in all languages. Scientific names are necessary because each language, and even different localities within one country, may have different common names for the same plant. For instance the scarlet-flowered weed *Anagallis arvensis* L. usually called "common pimpernel" or "scarlet pimpernel" in the United States, has 36 common names in English. In Germany the chosen common name for this species is "roter gauchheil" but there are also 84 other German names. In French it is "mouron" or any one of fourteen additional names, and the Dutch have a choice of 31. But in all countries it is known by the same scientific name.

The family names of most flowering plants end in the letters *–aceae* (pronounced ace-ee). Families are usually named after a genus in the family, as, for example, the Rosaceae. However, a few of the larger and more easily recognized families traditionally have not followed this pattern. Their names are based on older names for the family, and the shorter ending *–ae* is used. The Leguminosae (legume family), the Cruciferae (cruciform-flowered plants), and the Compositae (with inflorescences a composite head of flowers) are names of this type. For all of these families alternate names which follow the usual pattern are available and are coming into more frequent use. The alternate names are indicated in the family keys.

EDIBLE, MEDICINAL AND POISONOUS PLANTS

Edible Plants

In days past when the majority of people depended for fresh fruits and vegetables on what they could gather from the woods and fields, spring was the first time after a long, and possibly hungry, winter that green things could be gathered which would add diversity and vitamins, especially vitamin C, to the diet. Spring is also the time when the leaves of wild plants are the most tender and the least bitter, and therefore the most pleasant to eat. Consequently, leaves, flowers, and roots of many spring-flowering plants have been used for food. Later in the season, the fruits and seeds of some of these plants were also gathered. Some of these uses are listed in the descriptions in the keys.

Medicinal and Poisonous Plants

Other plants which have a place in this book have been used medicinally. Many times it is the plants which are poisonous in large doses that have been used to advantage when the doses are carefully controlled and the patient's reaction monitored. This medical use dates from the time when the only drugs available were made directly by boiling or grinding a plant rather than having the plant substances separated into component chemicals in the laboratory or made synthetically. Although some of the plants used medicinally have indeed been proven to have a curative effect for the condition they were used to treat, the crude extracts are no longer used by pharmacists. Extracts directly from plants often contain several active compounds, only one of which may be useful, while the others contribute undesirable side effects. Controlling the dosage of a drug prepared directly from a plant is difficult because the amount of drug in a plant part can vary with the age of the plant and the season of the year. For these reasons, although pharmacy owes much to drug plants and still obtains many of its raw materials from plants, unrefined plant extracts have nearly disappeared from the official lists of selected pharmacological substances, except as carriers for other drugs and as flavoring agents. One of the formerly popular medicinal plants which has been omitted from the United States Pharmacopeia is *Veratrum viride*. Until 1942 veratrum was listed as a source of powerful alkaloids used to control hypertension. Although effective, veratrum extract is difficult to use because the line between the effective dose and a dose which causes undesirable side effects, including heartburn, nausea and vomiting, is thin. New, similar, drugs which are easier to control and are considered more effective have replaced veratrum.

Bibliography

The following list includes a few of the books written about useful or poisonous plants. They will serve as starting points for readers who wish to know more.

Edible plants:

Fernald, M. L., and A. C. Kinsey. 1943. Edible wild plants of eastern North America. Idlewild Press, Cornwall-on-Hudson, New York.

Harrington, H. D. 1967. Edible native plants of the Rocky Mountains. University of New Mexico Press, Albuquerque, New Mexico.

McPherson, A., and S. McPherson. 1977. Wild food plants of Indiana and adjacent states. Indiana University Press, Bloomington, Indiana.

Medicinal Plants:

Krochmal, A. and C. 1973. A guide to the medicinal plants of the United States. Quadrangle/The New York Times Book Co., New York, New York.

Lewis, W. H., and M. P. F. Elvin-Lewis. 1977. Medical botany. John Wiley & Sons, New York, New York.

Poisonous plants:

Hardin, J. W., and J. M. Arena. 1969. Human poisoning from native and cultivated plants. Duke University Press, Durham, North Carolina.

Kingsbury, J. M. 1964. Poisonous plants of the United States and Canada. Prentice-Hall, Inc., Englewood Cliffs, New Jersey.

Kingsbury, J. M. 1965. Deadly harvest. Holt, Rinehart and Winston, Inc., New York, New York.

Youngken, Jr., H. W., and J. S. Karas. 1973. Typical poisonous plants. U.S. Department of Health, Education and Welfare, Bureau of Product Safety. For sale by the Superintendent of Documents, U.S. Government Printing Office, Washington, D. C. 20402.

Warning

It is important to emphasize that the medical uses and the edibility information given in this book are simply reports gathered from the literature. **NEITHER EATING THEM NOR CURING WITH THEM IS RECOMMENDED BY THEIR MENTION IN THIS BOOK.** These uses have not, for the most part, been tested by the authors. Anyone considering using wild plants for food or medicine is referred to books devoted solely to the subject. It *is* recommended, however, that anyone who nibbles plants in the wild pay particular attention to plants identified as poisonous in this book. Even plants not listed as poisonous cannot automatically be considered to be safe. While some wild plants can be eaten with no harmful, and sometimes even healthful effects, it is wise not to taste a plant part or fruit until identification is certain and the plant is known to be edible. It is well to remember also that the degree of individual reactions after either touching or eating a particular plant may vary with the sensitivity of the person, and in the case of internal use, with the size of the person and the amount of other food eaten at the same time which would dilute the toxic material.

If poisonous plants are eaten, first aid suggestions are to call a doctor or the nearest poison control center (usually located in hospitals) and induce vomiting. Take a complete sample of the plant so that it can be correctly identified.

Why are Plants Poisonous?

The poisonous properties in plants are one type of defense mechanism which protects plants from being eaten. The species of plants that have mechanisms which keep them from being grazed by animals, including insects and humans, are the ones which survive the best and reproduce the most. The defenses in some species are external ones like thorns or hairs. Other defenses are chemical. Some of these chemicals initially merely itch or sting or taste bitter and thereby simply discourage the animals. A larger dose of these chemicals may have a physiological effect and cause illness and finally even death. Either the animal learns not to eat members of that particular species or animals who persist in eating, die, and do not damage any other plants. Either way, this is an advantage to the plant species because the rest of the plants in the species live.

Of the plants in this book, 18% are listed as medicinal, 17% as edible, and 11% as poisonous, and many of these overlap!

COLLECTING AND PRESERVING

A collection of plants preserved by pressing and drying is an herbarium. Such a collection can be a useful reference or teaching tool or the documentation of an interesting hobby. There are a number of institutions worldwide which house large collections of herbarium specimens. In an herbarium, plant specimens from different parts of the world, different seasons of the year, and different periods of years can be compared side by side. An identified specimen may be used to identify other specimens by comparing the two.

Since many species of plants are threatened with extinction, new laws governing protected plants make complete collections difficult without a permit. In some states many of the most showy spring flowers are the ones which are protected. Until a person is familiar with state and federal laws concerning protected plants it is wise to adopt a "look but don't pick" attitude. For those who wish to keep a record of plants they have seen, check-off boxes are provided in the list of plants in the index.

Specimens of plants which are not considered threatened or endangered can be preserved as an herbarium. Plants are most easily preserved in the least space by pressing them flat and drying them. The flattened plants are then glued, pinned, or sewn to sheets of heavy paper and a label attached to the sheet which identifies the plant and the place where it was collected.

Collecting

A collector should be aware of local laws governing the collecting of plants and should be familiar with at least photographs of endangered and threatened plants so that one of them is not collected by mistake. (See p. 14 for more information on endangered plants).

The most valuable specimen is one which is complete and documented thoroughly (Fig. 12). A good specimen contains either the flower or fruit, since these are often the most

Figure 12

valuable parts for identification, and as much of the rest of the plant as can be bent into a V or N shape to fit onto a sheet of mounting paper. Roots are desirable if there are enough plants available so that digging one will not endanger the population.

For full documentation the specimen sheet should have a label (Fig. 13) which contains the following information:

★ the genus and species name of the plant and the author name
★ the location from which the plant was collected, beginning with the state, the county, and the nearest town, and containing enough information so that

someone else could find the collecttion locality again

★ a brief indication of the type of habitat in which the plant was growing, such as "oak woods" or "roadside"
★ a brief description of any part of the plant which was visible but not included on the specimen sheet, (for instance, the height of the plant if all of it could not fit on the sheet), and the color of the flower or fruit, since this is important in identification and the color often fades upon drying
★ the collector's name followed by the number the collector assigns sequentially to the specimen in the collecting book
★ the date of the collection.

PENNSYLVANIA
FLORA OF LEBANON COUNTY

Viola pubescens Ait.

PALMYRA, 0.5 miles north of Bindnagle's Church, along Swatara Creek. Moist woods on floodplain.

Flowers yellow.

Date: May 1, 1980
Coll: Susan Williams   no. 170

Figure 13

In the field, specimens may be pressed as soon as they are picked and the label data written on the press sheet or on a tag pressed with the plant. Specimens may also be collected and put into plastic bags to keep them somewhat fresh until they are pressed later. In this case each plant can be provided with a numbered slip of paper and the label information written under that number in a field notebook from which the labels can be typed later.

Pressing

Pressing a plant keeps it from wrinkling as it dries and allows the plant to be spread to show as many characters as possible. Equipment for pressing includes two pieces of plywood or lattice (the press), two pieces of rope or straps, sheets of corrugated cardboard with the corrugations running the same way in all sheets, blotting paper, and newspaper (Fig. 14). The standard size of press material used in American herbaria is 12 × 19 inches.

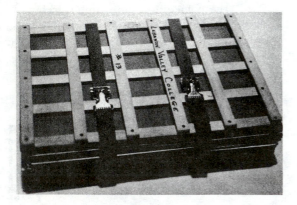

Figure 14

One of the pieces of lattice is used as the bottom frame of the press. On this bottom piece are stacked the specimens and alternate layers of cardboard and blotters. Plants are pressed between folds of half-sheets of newspaper. This newspaper is the folder in which the plant will be dried and stored until it is mounted. On it may be written the collection number of the plant pressed inside and any other data which will help to correlate the specimen with the data in the field notebook.

As the plants are pressed, the press material is built up sandwich-fashion, first a cardboard, then a blotter, then the newspaper with specimen inside, a blotter and a cardboard. A stack of pressed specimens will have each specimen sandwiched between two blotters and each set of blotters separated by a corrugated cardboard. The blotters absorb moisture rapidly from the plant and the corrugations in the cardboard allow air circulation through the press.

Initial positioning of the plant is important but it is still possible to change position the day after pressing, before the plant is completely dry. Since the pressed plant will be firmly attached to a sheet of paper, the parts should be arranged in order to show both sides when the specimen is mounted. At least one leaf should be turned to show the underside and if there are several flowers, one of those should be turned backwards also. At least one flower should be opened to show the number of parts. Large plants may be bent or folded to fit the press. Fleshy parts such as bulbs or fruits may be cut in half for drying.

When all of the collected plants have been pressed, the second lattice is added to the top of the stack and the straps pulled tight around each end of the press. If the plants were very succulent or moist at the time of pressing, the blotters may need to be replaced with dry ones during the next day. The press straps will need to be tightened daily.

Drying

Drying the plant quickly will result in a specimen that preserves more of the original color. Drying will be faster if a steady flow of air can be maintained through the corrugated cardboards of the press. This can be achieved by putting the press in a breezy place, by making a funnel of a sheet and putting the press in the large end with the corrugations running the same direction as the funnel and a hairdryer at the smaller end, or by supporting the press vertically over a light bulb which maintains a flow of air through the press. This last method is the most convenient and reliable. Support the press between two chairs, with the cardboard corrugations running vertically. Place a lamp with a 60-100 watt bulb underneath the press. The heat from the lamp will cause the air around it to rise through the press. The air need not be hot, simply moving. The lamp should not be so close to the press that it will be a fire hazard. Most of the spring flowers treated in this way will be dry in three to four days.

Mounting

When the specimen is thoroughly dry and a permanent label has been typed, the label and the specimen are attached to a sheet of stiff paper as a permanent mount. Standard herbarium size for paper is 11½ × 16½ inches. This size should be followed if the specimen is to be donated for scientific use. For a personal collection any size of paper is sufficient. The paper should be stiff enough to support the plant specimen and should have a high rag content so it will not crumble as it ages. The label should be glued to the lower right hand corner of the paper.

Glue is most frequently used to fasten the plant to the paper. Herbarium glue, mucilage or library paste is used. If the glue is squeezed from an applicator, apply glue to all parts of the plant—at the back of the flower, along the midrib of the leaf, and along the stem. Turn the plant over and drop it into position on the herbarium paper. Neatness and artistic positioning make a more attractive sheet.

Some herbarium glues are sprayed on the back of the specimen with a pump spray or brushed on with a brush. For delicate plants brush-on type glues are coated onto a piece of glass and the plant dropped onto the glass to pick up the glue and then dropped onto the paper.

While the glue dries it may be necessary to hold the plant parts in contact with the paper by means of metal washers used as weights. When the glue is dry, heavy stems or fruits should be fastened more securely using strips of gummed botanical tape or by making several loops with needle and thread or dental floss around the specimen and through the paper at strategic points.

The plant can also be sewn to the specimen sheet without gluing. If this is done, the back of the paper at the sewn spots should be reinforced with paper tape. Some herbaria in Europe pin specimens to the sheets, using straight pins pinned across the stems. White gummed botanical cloth tape can be used in the same manner. Adhesive tape is not recommended for mounting because eventually the adhesive creeps out from under the tape and sticks the specimen to the back of the specimen stacked above it.

Collecting and mounting supplies can be purchased from biological supply houses. Some sources of material are:

Carolina Biological Supply Company
Burlington, N.C. 27215

Carpenter/Offutt Paper Inc.
P.O. Box 3333
South San Francisco, CA 94080

Turtox/Cambosco
8200 Hoyne Avenue
Chicago, IL 60620

Ward's Natural Science Establishment, Inc.
P.O. Box 1712
Rochester, N.Y. 14603

Arrangement of the Collection

When the specimens are completely mounted and labelled they will be most useful if they are arranged in a sequence for easy reference. They can be arranged alphabetically by family and then alphabetically by genus within the family, or the families may be arranged following the sequence in this book. Species of the same genus are grouped together, alphabetically, in a *genus cover*. The genus cover used in herbaria is made of heavy manila paper which is folded in half. The genus name and the family name are printed on the outside.

If the sheets are carefully handled, kept dry and free from dust, not exposed to light for long periods of time, and protected from insects (by use of paradicholorobenzene crystals or other moth repellants), they will remain in good condition indefinitely. Specimens in many of the major herbaria are over one hundred years old. Those collected by Linnaeus and his colleagues are over 200 and are still valuable scientific specimens.

ENDANGERED SPECIES

In both the continental United States and Canada approximately 10% of the flora (and 50% of the Hawaiian flora) is considered rare and in danger of extinction from human threats.

One of the most common ways a plant species becomes extinct or is threatened with extinction is that its habitat is destroyed. Habitat destruction can come because land is converted for building or agricultural use or because low lying land is drained, filled, or flooded. The environment in which a plant grows can become polluted, or the pollinator on which it depends for reproduction may itself be threatened by insecticides or other environmental disruptions. The United States federal government has enacted the "Endangered Species Act of 1973" (Public Law 93-205, further amended in 1978 and 1979) which makes possible the legal protection of endangered and threatened species and also the "critical habitats" in which endangered species grow.

Plant species also become endangered by over-picking and digging, especially for commercial use. The Endangered Species Act also prohibits commerce in endangered and threatened species across state and national boundaries without a permit. The U.S. Forest Service and the U.S. Bureau of Land Management have complied with the spirit of the law by prohibiting picking of the endangered plants on lands under their jurisdiction. The U.S. National Park Service has prohibited the picking of any plants in national parks for many years. Many states and Canadian provinces have also enacted, or are considering, laws to protect species in their critical habitats or to prohibit picking or digging plants on state or provincial lands. In New York State for instance, 34 kinds of plants are protected from anyone who would "knowingly pick, pluck, sever, remove or carry away. . . ." Some of these 34 plants are single species; others are larger groups such as all orchids native to New York.

After the passage of the United States federal law in 1973, the Smithsonian Institution compiled a list of approximately 3000 species in danger of extinction nationwide. A revised list published in 1980 by the U.S. Fish and Wildlife Service included about 60 federally protected plants and again about 3000 species and varieties proposed, or under consideration for protection, as well as a list of some of the original 3000 that have been found to be not in danger and are no longer under review. Nearly every state now has a state list and some of these state lists are more extensive than the national list because they also include species which are in danger or rare within the state boundaries, although the species may be abundant in other states.

Two Canadian provinces, Ontario and New Brunswick, have laws which protect specific plants and some other provinces have wilderness area legislation which protects the habitats. The Canadian Botanical Society has proposed a list of rare and extinct plants for Canada which includes 1,364 plants. In addition, the National Museum of Natural Science is compiling lists of rare plants of the provinces and territories.

A collector should be thoroughly familiar with the protected plants of the local area and the rules governing them. Since fines can range up to $1,000 there is particular monetary interest in doing so. Until familiarity is gained it is better to "look but don't pick." Usually it is illegal to pick any plant from state, federal, or local park lands. Signs are usually posted at the boundaries to inform visitors. On private land permission of the land owner should be sought.

Federal and state laws and guidelines are in the process of being written and amended and are likely to be in the process of change for some time. It is not possible in this book to recount all of the state laws or even to indicate all of the plants which are protected by each state. Very few of the species in this book are under review for federal protection. These are indicated in the appropriate species descriptions by a star (★). If only a variety of a species is endangered, the star is followed by the abbreviation of the state in which the variety is possibly endangered.

In the United States, information about local laws and threatened plants can be obtained by writing to each state's Department of Natural Resources or Department of Conservation or to the Herbarium at the state university. The reference librarian at the local library can supply addresses for the state or regional Native Plant Society or Wild Flower Preservation Society as well as agency addresses from the *Conservation Directory* of the National Wildlife Federation. For information on Federally protected species write to the nearest regional office of the Department of the Interior, U.S. Fish and Wildlife Service:

Regional Director
U.S. Fish and Wildlife Service
Lloyd 500 Building, Suite 1692
500 E. Multnomah Street
Portland, Oregon 97232

Regional Director
U.S. Fish and Wildlife Service
500 Gold Avenue, S.W. (P.O. Box 1306)
Albuquerque, New Mexico 87103

Regional Director
U.S. Fish and Wildlife Service
Federal Building, Fort Snelling
Twin Cities, Minnesota 55111

Regional Director
U.S. Fish and Wildlife Service
Richard B. Russell Federal Building
75 Spring Street, S.W.
Atlanta, Georgia 30303

Regional Director
U.S. Fish and Wildlife Service
One Gateway Center, Suite 700
Newton Corner, Massachusetts 02158

Regional Director
U.S. Fish and Wildlife Service
P.O. Box 25486
Denver Federal Center
Denver, Colorado 80225

Area Director
U.S. Fish and Wildlife Service
1011 E. Tudor Road
Anchorage, Alaska 99503

In Canada, write to provincial departments of environment or wildlife management. Provincial lists of rare plants are available from the National Museum of Natural Science in Ottawa. Herbaria at the provincial universities are also a source of information.

Many wild plants have strict habitat requirements, which include a particular amount of water, temperature, soil type, amount of sunlight, pollinators, and associated plants. The more rare the plant the greater the probability that the growth requirements are very critical. Therefore attempts to protect species by digging them for cultivation in private gardens usually is not successful.

Collecting any plant carries with it the responsibility for the future of the species. Even with common plants it is not good practice to collect the only plant in a stand, or all of the plants from a population.

Care and concern for endangered plants should not be allowed to interfere with the average person's enjoyment of wildflowers in the springtime. An endangered plant is, by definition, rare, so that likelihood of coming across one and picking it is not very great. The greatest dangers of extinction come from habitat destruction and commercial over-exploitation rather than accidental picking by springtime botanists.

A very convenient and satisfying way of keeping track of the plants a person has identified is to follow the practice of the bird enthusiasts and simply check the plants off on a "life list," or in the index at the back of this book. The beauty of a flower is also enjoyed longer from a photograph than from a bouquet which soon goes limp. This author has recorded in the margin of the first edition of this book the date and place where she first saw the species in the wild, and in subsequent years, the dates of flowering.

MEASUREMENTS

The measurements given in this book are in English units with metric equivalents provided. I believe that the average American user of this key still has a better sense of the length of an inch or foot than a centimeter or a meter. The metric equivalents are as exact as moderate rounding will allow. This equivalency is maintained to allow the users of the key to develop a close sense of the length of a metric unit in terms of the more familiar English units.

In the descriptions in this book conversions have been made from both English to metric and vice versa. These data have been taken from the previous edition and compiled from various technical works.

A single plant is really a member of a population of plants in a particular place and a member of a group of populations which make up a species. Within a species there is a certain natural range of variation in the shape and size of plant parts. For this reason the measurements in the keys are usu-

ally given for the range of variation. The smaller the size of a part the more critical the measurement must be. A quarter inch or a few millimeters may make the difference between placing the plant in one species or another. For this reason it is important for the equivalent measurements to be as exact as can be measured. Larger measurements such as leaf width or plant height are less critical and less exact. A plant may be usually 3 feet tall but this 3 feet measurement has a latitude of several inches on either side. An inch equals 2.54 cm so that the exact metric equivalent of 3 feet, 91.44 cm, actually means 91 cm plus or minus 6 or 8 cm and the significance of a rounded figure of 90 cm is virtually the same as the 91 cm equivalent. Although when one is dealing with a length of 90 cm the extra 1 cm is not significant, the exact equivalents have been maintained in this book for the reasons stated above.

How to Use the Keys

A key is a tool for identification. By the process of elimination it leads step by step to the identification of the plant. A key is somewhat like a map to a road which constantly divides in equal forks. At each fork the driver must make a decision whether to go left or right, based on information found at that fork. After a series of correct decisions the driver arrives at the destination. In the case of this key, the destination is the identification of the plant.

At each step in the key there are two short descriptive statements. Each pair of statements is called a *couplet*. Each couplet is numbered and each half of the couplet is designated *a* or *b*. At the end of each half of the couplet there is either a name or another number. The name is the identification of the flower, the end of the line. The number is the number of another couplet, the next in the line of clues.

To use the key, read both halves of the first couplet. Be sure to read all of each couplet. Compare the descriptions with the plant to be identified. Decide which half of the couplet fits the plant at hand and look to the end of the line for the number of the next clue.

If neither description fits completely, a mistake was made in a previous choice. In this case, go back to a choice that was debatable and try the other route.

Once an identification is reached it should be checked further by reading the descriptive paragraph, comparing the illustration, and comparing the map of the range with the actual locality of the plant.

This key is divided into two parts. The first part identifies plants to the family level. The second part contains keys for the individual families, arranged in order of their botanical relationship. An alphabetical index of families is on the last page of this book.

The keys to families include families of the common spring flowering plants east of the Rocky Mountains which are herbaceous. Grasses and sedges are included in the family key but because they are specialized groups they are not keyed further here. For specific identification see Pohl, *How to Know the Grasses*.

Keys to Major Groups and Families

This key applies only to spring-flowering species within North America east of the Rocky Mountains. It is not arranged to show relationships among families. The family relationships are indicated by the sequence of the family keys.

Angiospermae Class Key

1a Leaves usually parallel-veined (a); flowering parts usually in 3's (c); stems hollow or with scattered vascular bundles (b); seeds usually with one cotyledon (Fig. 15) .... page 30, MONOCOTYLEDONS

Figure 15

1b Leaves usually net-veined (a); parts of the flower mostly in 4's or 5's (c); stems with a ring of vascular bundles surrounding the pith (b), or woody with outer bark (woody plants not included in this book); seeds usually with two cotyledons (Fig. 16) .... page 71, DICOTYLEDONS

Figure 16

Family Key—Monocotyledons

1a Perianth (sepals and petals) absent or scale-like and inconspicuous 2

1b Perianth present, herbaceous or colored, not scale-like (Fig. 17) 4

Figure 17

2a Inflorescence of flowers on a fleshy spike (*spadix*) accompanied by an enlarged bract (*spathe*, Fig. 18) .... page 31, ARACEAE

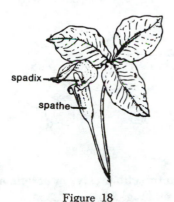

Figure 18

2b Inflorescence a spikelet (Fig. 19), the spikelets often grouped together into a larger inflorescence of various forms; leaves grasslike 3

Figure 19

3a Flowers (pistils and stamens) enclosed in a pair of bracts, stems usually hollow, round in cross section .... GRAMINEAE*

3b Flowers in the axil of a single bract, stems usually solid, often triangular in cross section .... CYPERACEAE*

4a Plants growing upon other plants (epiphytes), hanging from trees or attached to the trunk and branches .... page 36, BROMELIACEAE

4b Plants terrestrial or aquatic 5

5a Pistils many, in a head (flowers or plants may be unisexual); leaves basal .... page 30, ALISMATACEAE

5b Pistil one 6

6a Stamens and pistil united into a column; flowers bilaterally symmetrical (zygomorphic) (Fig. 20) .... page 60, ORCHIDACEAE

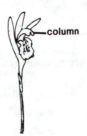

Figure 20

6b Stamens and pistil free from each other 7

7a Plant a floating aquatic .... page 36, PONTEDERIACEAE

*Gramineae and Cyperaceae are not treated in this book. For identification of these specialized families see Pohl's *How to Know the Grasses.*

7b Plant terrestrial or growing in marshes or along streams and lake shores **8**

8a Stamens 6 .... **9**

8b Stamens 3 ....
.... page 55, **IRIDACEAE**

9a Ovary superior (Fig. 21) **10**

Figure 21

9b Ovary inferior (Fig. 22) ....
.... page 54, **AMARYLLIDACEAE**

Figure 22

10a Plant with soft annual leaves **11**

10b Plant with coarse, fibrous, perennial leaves .... page 55, **AGAVACEAE**

11a Leaves expanded at the base into a sheath surrounding the stem ....
.... page 34, **COMMELINACEAE**

11b Leaves not expanded into a sheath at the base .... page 37, **LILIACEAE**

Family Key—Dicotyledons

1a Perianth (sepals and petals) absent; involucre petal-like (a); ovary of female flower protruding on a stalk (b); juice milky (Fig. 23) ....
.... page 145, **EUPHORBIACEAE**

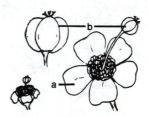

Figure 23

1b Perianth (either calyx or corolla or both) present (Fig. 24) **2**

Figure 24

2a Pistils usually 3-many, sometimes as few as 2 in some flowers (Fig. 25) **3**

Figure 25

2b Pistils 1 or 2 (Fig. 26) 4
Individual flowers may be compacted into a flower-like head, see couplet 36a or p. 212.

Figure 26

3a Sepals 5; petals 5; stamens 5-numerous; flowers with a hypanthium (Fig. 27) .... page 126, ROSACEAE

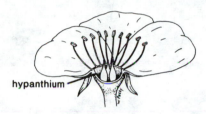

hypanthium

Figure 27

3b Sepals 3-many; petals 3-many or absent; stamens numerous; flowers without a hypanthium .... page 85, RANUNCULACEAE

4a Corolla absent; sepals sometimes showy and petal-like (Fig. 28) 5

Figure 28

4b Corolla present (Fig. 29) 11

Figure 29

5a Ovary inferior (Fig. 30) 6

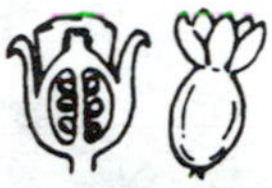

Figure 30

5b Ovary superior (in Nyctaginaceae the calyx is constricted so that the ovary appears inferior) (Fig. 31) 8

Figure 31

6a Stamens 12; calyx lobes 3, magenta; locules 6 .... page 71, ARISTOLOCHIACEAE

6b Stamens 4-8, growing from a disk at the center of the flower; calyx 4- or 5-parted; locule 1 7

7a Plant creeping; calyx 4-lobed; stamens 4-8 .... page 123, SAXIFRAGACEAE (*Chrysosplenium*)

7b Plant erect; sepals 5, greenish white, petal-like; stamens 5 .... page 71, SANTALACEAE

8a Leaves alternate (Fig. 32) 9

Figure 32

8b Leaves opposite or whorled (Fig. 33) 10

Figure 33

9a Stipules present, forming sheaths around the stem (Fig. 34); calyx 3- to 6-parted; flowers perfect or unisexual .... page 72, POLYGONACEAE

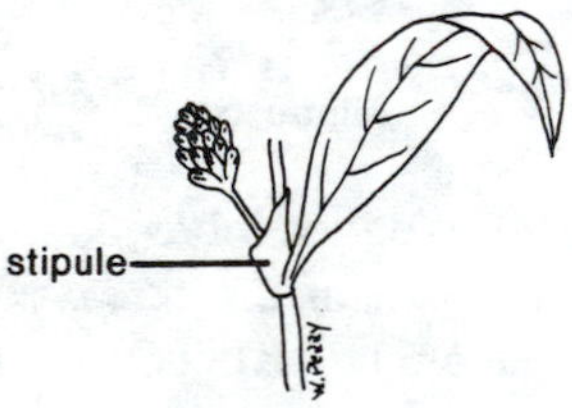

stipule—
Figure 34

9b Stipules absent; lower leaves reduced to scales; calyx-like bracts 4; flowers either staminate or pistillate .... page 147, BUXACEAE

10a Flowers rose or purplish; sepals united; leaves opposite .... page 75, NYCTAGINACEAE

10b Flowers greenish white; sepals separate; leaves whorled .... page 75, AIZOACEAE

11a (4) Petals all separate (a), or in zygomorphic (bilaterally symmetrical) flowers several of the petals united (b) (Fig. 35) 12

Figure 35

11b Petals all united (Fig. 36) 36

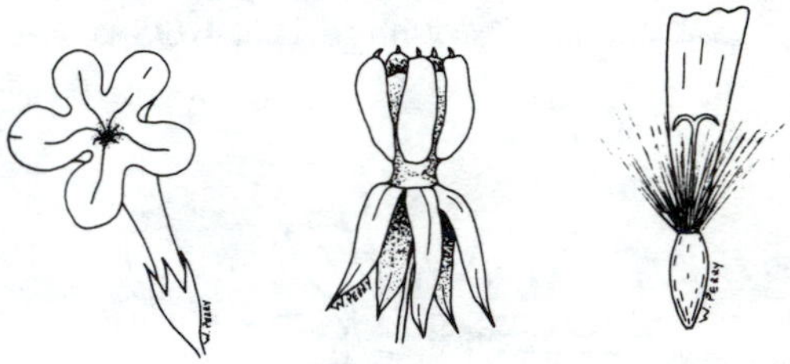

Figure 36

12a Ovary superior (a) or partly inferior (b) (Fig. 37) 13

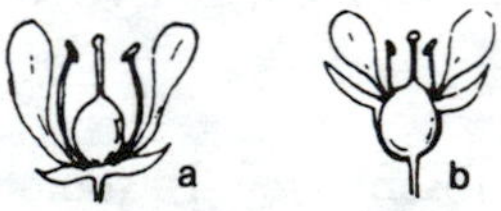

Figure 37

12b Ovary completely inferior (Fig. 38) 33

Figure 38

13a Stamens more than 10 14

13b Stamens 10 or fewer 20

14a Plants aquatic or growing in bogs 15

14b Plants terrestrial 16

15a Leaves flat, usually floating; plant aquatic .... page 84, NYMPHAEACEAE

15b Leaves tubular; plants of bogs ....
.... page 122, SARRACENIACEAE

16a Inflorescence a long, terminal raceme; carpel 1; fruit berry-like ....
page 94, RANUNCULACEAE (*Actaea*)

16b Inflorescence axillary, or if terminal, not a raceme, or flowers solitary 17

17a Stamens united into a column around the single pistil; carpels many, in a ring ......
.... page 148, MALVACEAE

17b Stamens free; carpels 1-several in a 1 locular pistil (Fig. 39) 18

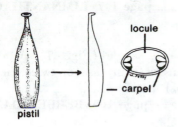

Figure 39

18a Plants with colored juice; sepals 2 ....
.... page 103, PAPAVERACEAE

18b Plants without colored juice; sepals 5 or 6 19

19a Leaves lobed, peltate, in a terminal pair on flowering stems; flowers solitary; sepals 6; petals 6-9, white ....
.... page 103, BERBERIDACEAE (*Podophyllum*)

19b Leaves entire, alternate; flowers terminal or axillary, of 2 types, large with 5 yellow petals or small and apetalous; sepals 5 .... page 148, CISTACEAE

20a (13) Corolla strongly zygomorphic (Fig. 40) 21

Figure 40

20b Corolla actinomorphic (radially symmetrical) or nearly so (Fig. 41) ....
.... 24

Figure 41

21a Leaves simple, entire or lobed (Fig. 42)
.... 22

Figure 42

21b Leaves compound (Fig. 43) ....
.... 23

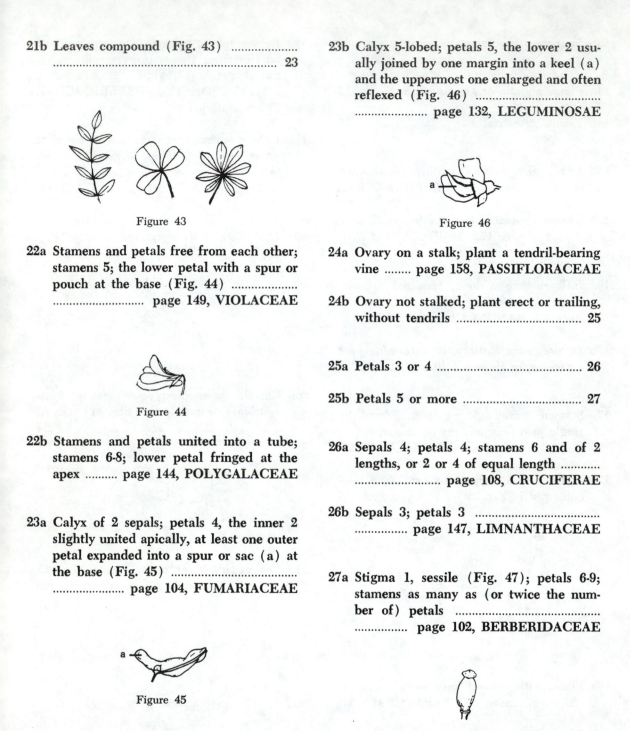

Figure 43

22a Stamens and petals free from each other; stamens 5; the lower petal with a spur or pouch at the base (Fig. 44) ....
.... page 149, VIOLACEAE

Figure 44

22b Stamens and petals united into a tube; stamens 6-8; lower petal fringed at the apex .... page 144, POLYGALACEAE

23a Calyx of 2 sepals; petals 4, the inner 2 slightly united apically, at least one outer petal expanded into a spur or sac (a) at the base (Fig. 45) ....
.... page 104, FUMARIACEAE

Figure 45

23b Calyx 5-lobed; petals 5, the lower 2 usually joined by one margin into a keel (a) and the uppermost one enlarged and often reflexed (Fig. 46) ....
.... page 132, LEGUMINOSAE

Figure 46

24a Ovary on a stalk; plant a tendril-bearing vine .... page 158, PASSIFLORACEAE

24b Ovary not stalked; plant erect or trailing, without tendrils 25

25a Petals 3 or 4 26

25b Petals 5 or more 27

26a Sepals 4; petals 4; stamens 6 and of 2 lengths, or 2 or 4 of equal length ....
.... page 108, CRUCIFERAE

26b Sepals 3; petals 3 ....
.... page 147, LIMNANTHACEAE

27a Stigma 1, sessile (Fig. 47); petals 6-9; stamens as many as (or twice the number of) petals ....
.... page 102, BERBERIDACEAE

Figure 47

27b Stigmas 2-5, on a style (Fig. 48); petals 5, sometimes deeply notched; stamens 2-10 28

Figure 48

28a Sepals 2; petals 5; stamens same number as the petals and opposite them (Fig. 49); stem leaves 2, opposite ....
.... page 76, PORTULACACEAE

Figure 49

28b Sepals 4-5; petals 5; stamens not the same number as the petals, or if the same number as the petals, alternate with them (Fig. 50) 29

Figure 50

29a Stigmas 2 ....
.... page 123, SAXIFRAGACEAE

29b Stigmas 3-5 30

30a Leaves opposite (Fig. 51) 31

Figure 51

30b Leaves alternate (a), whorled (b) or basal (c) (Fig. 52) 32

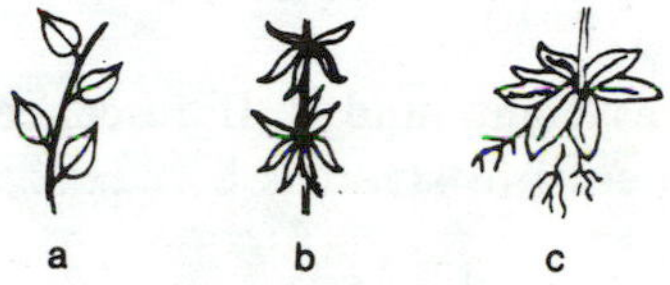

Figure 52

31a Leaf blades entire ....
.... page 77, CARYOPHYLLACEAE

31b Leaf blades dissected or deeply lobed ....
.... page 142, GERANIACEAE

32a Leaves trifoliolately compound, alternate or basal (Fig. 53) ....
.... page 141, OXALIDACEAE

Figure 53

32b Leaves simple, mostly in whorls of 3 ......
.... page 123, CRASSULACEAE

33a (12) Inflorescence a head-like cluster surrounded by 4 large petal-like whitish bracts ....
.... page 166, CORNACEAE

33b Inflorescence an umbel (Fig. 54) or flower solitary 34

Figure 54

34a Leaves simple; flowers solitary in the axils of upper leaves; petals 4; stamens .. 8 .... **page 158, ONAGRACEAE**

34b Leaves compound; inflorescence an umbel; petals 5; stamens 5 35

35a Petioles sheathing the stem; leaves aromatic; stems hollow .... **page 160, UMBELLIFERAE**

35b Petioles not sheathing the stem; leaves not aromatic; stems solid .... **page 159, ARALIACEAE**

36a (11) Inflorescence a head surrounded by 1 or more rows (a) of bracts (involucre) (Fig. 55) .... **page 212, COMPOSITAE**

Figure 55

36b Inflorescence not a head surrounded by an involucre 37

37a Ovary superior (Fig. 56) 38

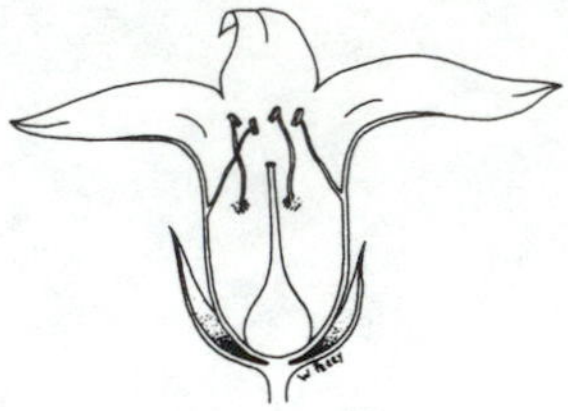

Figure 56

37b Ovary inferior (Fig. 57) 54

Figure 57

38a Flowers actinomorphic (radially symmetrical) (Fig. 58) 39

Figure 58

38b Flowers strongly or weakly zygomorphic (bilaterally symmetrical) (Fig. 59) 51

Figure 59

39a Inflorescence a crowded spike at the top of a scape; flowers small, flower parts in 4's; leaves all basal .... page 203, PLANTAGINACEAE

39b Inflorescence not crowded in a spike; flower parts in 5's; leaves on the stem, or if basal then the flowers large and showy 40

40a Flower highly modified, with a corona (a), and with anthers and stigma united; pollen in waxy masses (Fig. 60) .... page 173, ASCLEPIADACEAE

Figure 60

40b Flower not highly modified, corona lacking; pollen powdery 41

41a Ovaries 2 (Fig. 61); style 1; sap milky .... page 172, APOCYNACEAE

Figure 61

41b Ovary 1, deeply lobed in some families (Fig. 62) 42

Figure 62

42a Stamens more in number than the corolla lobes .... page 166, ERICACEAE

42b Stamens the same number as the corolla lobes 43

43a Ovary deeply 4-lobed; immature inflorescences more or less coiled; corolla tube with pubescent lines, folds, or crest-like appendages in the throat .... page 185, BORAGINACEAE

43b Ovary not deeply lobed 44

44a Stamens opposite the corolla lobes .... page 167, PRIMULACEAE

44b Stamens alternate with the corolla lobes 45

45a At least the lower leaves alternate, simple, entire or lobed or with a few small basal leaflet-like lobes (Fig. 63) 46

Figure 63

45b Leaves opposite (a), whorled (b), or compound and alternate (c) (Fig. 64) 48

Figure 64

46a Sepals separate, accompanied by 2 bracts; flowers white or pinkish; plant a slender, trailing vine .... page 176, CONVOLVULACEAE

46b Sepals united at the base; plants erect, or if trailing, with purple flowers 47

47a Style branched at the tip; leaves acutely or deeply lobed .... page 181, HYDROPHYLLACEAE

47b Style unbranched; leaves lanceolate and toothed or ovate with a few small basal lobes .... page 194, SOLANACEAE

48a Plants growing in marshes; leaves trifoliolate .... page 171, MENYANTHACEAE

48b Plants terrestrial 49

49a Stipular lines present between the leaf bases (Fig. 65-a) .... page 170, LOGANIACEAE

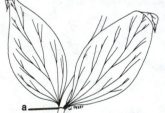

Figure 65

49b Stipular lines or stipules absent 50

50a Calyx 5-lobed; corolla 5-lobed; stigmas 3; leaves or leaflets oval to filiform .... page 177, POLEMONIACEAE

50b Calyx lobes 2, leafy; corolla 4-lobed; stigmas 2; lower leaves reduced to scales .... page 171, GENTIANACEAE

51a (38) Plants without chlorophyll; leaves alternate, scale-like .... page 202, OROBANCHACEAE

51b Plants with chlorophyll; leaves not scale-like 52

52a Corolla nearly radially symmetrical, tubular with a spreading limb (Fig. 66) .... page 189, VERBENACEAE

Figure 66

52b Corolla bilaterally symmetrical (Fig. 67) 53

Figure 67

53a Stems square in cross section; flowers in whorls on the stem (the whorls sometimes crowded into a head or spike) or flowers solitary; ovary 4-lobed (Fig. 68); style arises from under the base of the ovary; ovules 4; foliage aromatic .... page 190, LABIATAE

Figure 68

53b Stems rounded in cross section; flowers in racemes or solitary; ovary not lobed (Fig. 69), 2-celled; style terminal; ovules numerous .... page 195, SCROPHULARIACEAE

Figure 69

54a (37) Leaves alternate; corolla campanulate (Fig. 70) .... page 211, CAMPANULACEAE

Figure 70

54b Leaves opposite or whorled; corolla tube cylindrical or funnelform, long or short, the limb (a) either erect or spreading (Fig. 71) 55

Figure 71

55a Plants coarse, 16 inches (40 cm) or more tall; leaves without stipules; corolla lobes and stamens 5 .... page 210, CAPRIFOLIACEAE

55b Plants less than 12 inches (30 cm) tall or trailing; leaves either opposite and with stipules or whorled; corolla lobes and stamens usually 4 .... page 205, RUBIACEAE

ALISMATACEAE

WATER PLANTAIN FAMILY

1a Leaves linear to ovate, grass-like or spoon-shaped with a long petiole and ovate blade; stamens 12-20; filaments with scales (Fig. 72) ....
...... **NARROW-LEAVED SAGITTARIA**, *Sagittaria graminea* Michx.

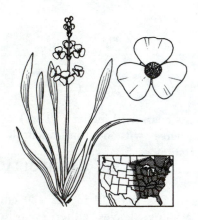

Figure 72

Rooted in mud or shallow water. The plant is either erect or submersed. When the plant grows under water, the leaves do not form blades, but the petiole expands and becomes leaf-like. The leaf blades are 1½ to 12 inches (4-30 cm) long and up to 4 inches (10 cm) wide. The flower stalk is 1 1/8 to 20 inches (3-50 cm) tall. The white flowers of both this and the following species have 3 sepals and 3 deciduous petals. Flowers in whorls higher on the inflorescence may have stamens only and the lower flowers may have pistils only.

1b Leaves sagittate; stamens 25-40; filaments smooth (Fig. 73) ....
.... **COMMON ARROWHEAD**, *Sagittaria latifolia* Willd.

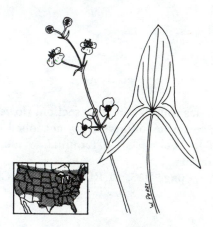

Figure 73

In swamps, ponds, and on stream banks. These plants grow erect with a flowering stalk of 4 to 47 inches (10-120 cm) tall. The leaves are very variable; the lobes and "tails" may be either broad or very narrow, the whole leaf ranging from 1 to 20 inches (2-25 cm) wide. The base of the plant bears edible starch-containing tubers that were roasted by the North American Indians. Another common name is Wapato. A pubescent variety, *S. latifolia* var. *pubescens* (Muhl.) J. G. Smith, occurs along the east coast.

ARACEAE

ARUM FAMILY

1a Fleshy spadix accompanied by a broad bract (*spathe*) **2**

1b Fleshy spadix accompanied by a long and narrow spathe (Fig. 74) ....
.... **SWEET FLAG,**
Acorus calamus L.

Figure 74

Wet ground. A plant 1 to 4 feet (31-122 cm) high with stiff, linear leaves, much like those of cattails. The tiny yellowish flowers are crowded on a cylindric spadix which is attached laterally to a stem that resembles the leaves. The horizontal rhizome is spicy-aromatic and furnishes the drug calamus used in medicine and flavorings as a "bitter," and in perfumery. The rhizome can be candied by boiling in sugar syrup for 2 to 3 days.

2a Spadix spherical or ovoid **3**

2b Spadix elongate, enveloped in a thin spathe **4**

3a Spadix (a) spherical, enveloped in a fleshy, hood-like spathe (Fig. 75) ....
.... **SKUNK CABBAGE,**
Symplocarpus foetidus (L.) Nutt.

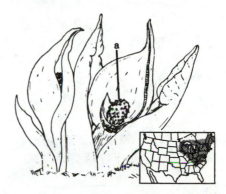

Figure 75

In bogs and moist areas. This is the earliest spring flower, often appearing in February. The spathe is about 6 inches (15 cm) high and brownish purple to greenish yellow, sometimes mottled with both colors. The fetid odor, a combination of skunk and cabbage, is evident if the spathe is picked or otherwise injured. A sweeter odor is produced as the plant heats up as a result of biochemical reactions in the spadix. This odor attracts pollinating insects. The heat spreads the odor, encourages pollinating insects to stay longer within the spathe, and also allows the skunk cabbage to melt its way upward through the ice and snow. The large, oval leaves, 1 to 2 feet (30-61 cm) long, appear after the flowers and are conspicuous all summer. Leaves and roots have been eaten but care must be taken in preparation to remove the irritating calcium oxalate crystals. In the same habitat, *Veratrum viride* (Fig. 115) has somewhat similar leaves and is highly poisonous when eaten.

3b Spadix ovoid, with an open, spreading, petal-like spathe white on the inside (Fig. 76) .... **WILD CALLA,** *Calla palustris* L.

4a Leaves simple, sagittate; flowers covering spadix (Fig. 77) .... **GREEN ARROW-ARUM,** *Peltandra virginica* (L.) Schott

Figure 76

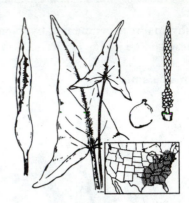

Figure 77

A slender bog plant 5 to 12 inches (13-30 cm) high. The spathe is ovate with a thin elongate point; the upper surface is greenish white, the lower surface green. The spathe may attain a length of 2 inches (5 cm) or more; the heart-shaped leaves have a width up to 4 inches (10 cm). The flowers contain both stamens and pistils. The red berries form a cluster similar to that of Jack-in-the-pulpit.

A coarse marsh plant 1 to 1 1/2 feet (30-46 cm) high often growing in water. The sagittate leaves are oblong to triangular and are sometimes over 20 inches (51 cm) in length. The pointed green spathe is 3 to 7 inches (7.5-18 cm) long; in the spadix the staminate flowers are above and the pistillate flowers are below.

4b Leaves compound; flowers at base of spadix only 5

5a Leaves mostly 2, divided into 3 leaflets; spadix much shorter than spathe (Fig. 78) .... **JACK-IN-THE-PULPIT,** *Arisaema triphyllum* (L.) Schott

5b Leaf usually solitary, pedately divided into 7 to 15 leaflets; spadix tapering to a long point beyond the spathe (Fig. 79) .... **GREEN-DRAGON,** *Arisaema dracontium* (L.) Schott

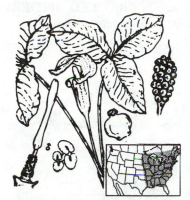

Figure 78

Figure 79

Damp woods. The plant reaches 1 to 2 1/2 feet (30-76 cm) high. The spathe (the "pulpit") is green and often striped purplish brown. The spadix ("Jack") is green or maroon. The underground corm is edible when boiled, giving rise to the common name Indian Turnip. The raw corm contains crystals of calcium oxalate, as do most members of this family, which injure the membranes of the mouth and throat, causing burning and swelling. The fruits are a cluster of bright orange-red berries.

Many common house plants such as *Philodendron* and *Dieffenbachia* are also in this family—and also contain calcium oxalate crystals.

Rich woods. Closely related to Jack-in-the-pulpit but not as common or conspicuous. The spadix is prolonged beyond the greenish, pointed spathe in what is probably the "dragon's tail." The plant is 1-3 feet (30-91 cm) high. In the fall it bears a cluster of bright red-orange berries.

COMMELINACEAE

SPIDERWORT FAMILY

Typical of this family is a perianth of 3 persistent herbaceous sepals and 3 ephemeral petals (pink, white, or blue); stamens 6, the 3 upper of them often with sterile, cross-shaped anthers.

1a Bracts under flowers heart shaped but folded; corolla irregular, one petal smaller; fertile stamens 3, filaments not hairy (Fig. 80) .... DAYFLOWER, *Commelina erecta* L.

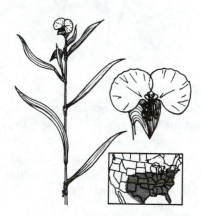

Figure 80

Sandy soil. Plant up to 3 feet (91 cm) tall. The leaves are linear or lanceolate. A group of flowers is surrounded by a folded bract, only one flower blooming at a time and lasting less than a day. The two upper petals are blue, the lower petal is smaller and white. The smooth brown seeds are eaten by birds. This species blooms from May into fall.

1b Bracts under flowers 2, elongate, resembling the leaves; corolla regular, all 3 petals of equal size; fertile stamens 6 (genus **TRADESCANTIA**) 2

2a Sepals glabrous or with non-glandular hairs only 3

2b Sepals with glandular hairs (Fig. 81) .... **BRACTED SPIDERWORT**, *Tradescantia bracteata* Small

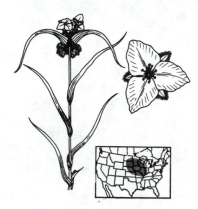

Figure 81

This midwestern species is about 15 inches (38 cm) tall with narrow and recurved leaves. The bracts are as large as the leaves. These flowers with rose or blue petals bloom in April and May. The hairs on the sepals and pedicels are both glandular and non-glandular.

Another similar rose-to-blue-flowered midwestern species, PRAIRIE SPIDERWORT, *Tradescantia occidentalis* (Britt.) Smyth, has only glandular hairs on the sepals and pedicels. This species blooms somewhat later over most of its range. It grows from western Wisconsin and Minnesota to Utah and Texas.

3a Pedicels and sepals with hairs all over (Fig. 82) .... **VIRGINIA SPIDERWORT,** *Tradescantia virginiana* L.

3b Pedicels without hairs; sepals without hairs or with hairs only in tufts at base and tip (Fig. 83) .... **OHIO SPIDERWORT,** *Tradescantia ohiensis* Raf.

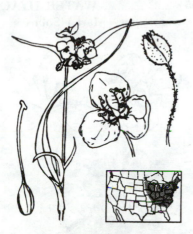

Figure 82

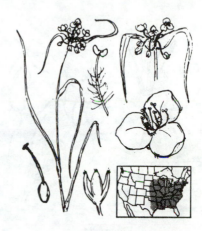

Figure 83

This spiderwort grows usually 1 to 1 1/2 feet (30-46 cm) tall and has bright green leaves. The stems and leaves may be either glabrous or with a fine pubescence. The flowers, usually blue, may be also rose, violet, or white. It is found in moist soil in woods and meadows and is frequently cultivated.

WANDERING JEW, *Tradescantia fluminensis Vell.,* a creeping plant with ovate leaves and white flowers, native to South America, is frequently cultivated and grows escaped from cultivation on the coastal plain from North Carolina to Florida and along the Gulf of Mexico.

This member of the genus, common in many places, may be recognized by its glaucous stems and leaves. The leaves are quite narrow and the bracts turn downward and may be even longer than the leaves. The flowers are bright blue, sometimes rose, rarely white. It grows in sandy soil in prairies and along railroads and attains a height of 1 to 3 feet (30-91 cm).

BROMELIACEAE

PINEAPPLE FAMILY

(Fig. 84) .... **SPANISH MOSS,**
Tillandsia usneoides (L.) L.

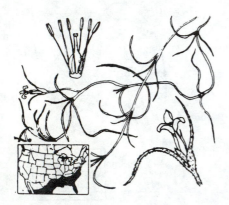

Figure 84

Plants moss-like, hanging in long silvery-gray tufts from tree branches. The scales on the thread-like leaves trap water and dust-borne nutrients for the plant. The small flowers are greenish yellow. This epiphyte is a characteristic feature of the southeastern states. It drapes trees, shrubs, telephone wires and fences. Plants have been used as mattress and upholstery stuffing, but in the wild they are often homes for chiggers. In South America and in South Carolina it has been used for its medicinal qualities. The Pineapple is in the same family.

PONTEDERIACEAE

PICKEREL WEED FAMILY

(Fig. 85) .... **WATER HYACINTH,**
Eichornia crassipes (Mart.) Solms

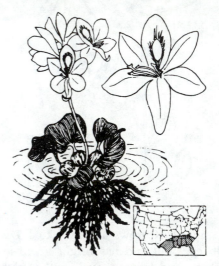

Figure 85

The leaf petioles are inflated into air bladders so that the entire plant floats wherever the water is at all deep. The leaves are oval or kidney-shaped and vary greatly in size so that the plants may attain a height of 3 feet (91 cm) or more. The purplish-blue flowers, an inch (2.5 cm) or more across, are borne in showy spikes or panicles. Patches frequently cover several acres and are unrivalled for their display. This plant was introduced from South America and has become a pest, choking many of the waterways of the South.

LILIACEAE

LILY FAMILY

1a Plants dioecious (two plants, staminate flowers on one plant, pistillate flowers on the other); inflorescence spicate-racemose (Fig. 86) .... **BLAZING-STAR**, *Chamaelirium luteum* (**L.**) **Gray**

Figure 86

The leaves in the basal rosette are obovate to spatulate and 3 to 6 inches (7.5-15 cm) long; the leaves on the stem become elliptic and gradually smaller upwards. The white flowers mature into a fruit that is an oblong capsule. Another name for this plant is Devil's-bit. The rhizome and roots have been used as a tonic against intestinal worms.

1b Plants monoecious (stamens and pistils on the same plant or in the same flower) **2**

2a True leaves scale-like; apparent leaves (really branches) filiform (Fig. 87) .... **ASPARAGUS**, *Asparagus officinalis* **L.**

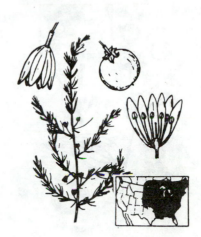

Figure 87

Asparagus was introduced from Europe and has escaped from cultivation to hedgerows and fields widely throughout the United States. The plant 3-7 feet (1-2 m) high bears small greenish-yellow, lily-like flowers and, later, red berries. The young shoot, which is the edible part, grows amazingly fast—sometimes several inches a day.

2b Leaves broader, not scale-like or thread-like **3**

3a Flowers 4-parted (Fig. 88) .... CANADA MAYFLOWER, *Maianthemum canadense* Desf.

Figure 88

A glabrous or pubescent plant with usually two leaves, though occasionally one or three, and attaining a height of 3 to 7 inches (7.5-18 cm). The leaves are ovate, small, and usually heart shaped at the base. The white flowers are borne in terminal racemes and produce small pale red berries which are usually speckled with darker red. It is also known as Wild Lily-of-the-valley.

3b Flowers 6-parted, all perianth parts alike (*tepals*) or with the 3 sepals unlike the 3 petals 4

4a Perianth segments united, 6 lobed (Fig. 89) 5

Figure 89

4b Perianth segments distinct, separate to the base or nearly so (Fig. 90) 8

Figure 90

5a Leaves alternate along the stem, the flowers cylindrical, greenish, axillary 6

5b Leaves basal or nearly so 7

6a Leaves without petioles, glabrous, leaf base clasping about half way around the stem (Fig. 91) .... GREAT SOLOMON'S-SEAL *Polygonatum biflorum* (Walt.) Ell.

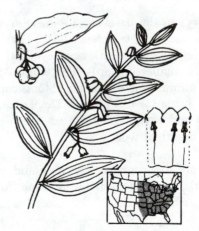

Figure 91

This plant gets its common name from the scar (the "seal") left on the rhizome where the leafy stalks of previous years were attached. The stems are 2 to 8 feet (60-244 cm) high and can be erect or arching. The elliptic to broadly oval leaves are smooth on both sides, sessile, and slightly clasping. Despite the scientific name the greenish-white flowers can be borne in groups of one to fifteen on a peduncle, although 2 to 4 flowers is usual. The berries are blue black and usually covered with a bloom. Although the berries cause "purging," the young shoots may be eaten like asparagus. The Indians used the rhizomes as a source of starch, and to prepare a tonic.

6b Leaves narrowed to a very short petiole, hairy on the underside in lines along the veins (Fig. 92) ....
.... **HAIRY SOLOMON'S-SEAL,**
Polygonatum pubescens (Willd.) Pursh

Figure 92

Woods. This Solomon's-seal is not as large as *Polygonatum biflorum.* It grows only 1 to 3 feet (30-91 cm) high. The leaves can be elliptic to broadly oval. The greenish-yellow flowers are about 3/8 inch (1 cm) long and are borne 1 to 4 (usually 2) to a peduncle. Species of this genus also grow in Europe, China, and Japan.

7a Leaves narrow, flowers usually blue (Fig. 93) .... **GRAPE HYACINTH,**
Muscari botryoides (L.) P. Mill.

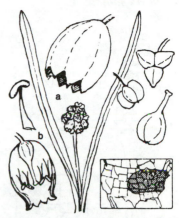

Figure 93

The Grape Hyacinth (a) has narrow leaves and a long scape all arising from a small bulb. It attains a height of four to ten inches (10-25 cm) along roadsides and in fields where it has escaped from cultivation, being native of Europe. The flowers are usually purplish-blue but occasionally white. The rather large green, six-seeded capsules are much more conspicuous than the flowers.

STARCH GRAPE-HYACINTH, *Muscari atlanticum* Boiss. & Reut. (Fig. 93b.) differs from the above in the flowers being elongated and the leaves more slender, nearly cylindrical and recurving downward. It too has escaped from cultivation.

7b Leaves elliptical, pointed at the apex, usually 2 or 3; flowers white with the lobes recurved (Fig. 94) .... LILY-OF-THE-VALLEY, *Convallaria majalis* L.

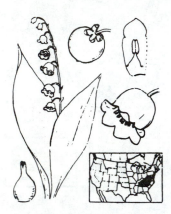

Figure 94

These fragrant flowers have been introduced from Europe and have escaped from gardens. It is interesting that the very same species grows wild in high mountains from Virginia to South Carolina. This sort of thing occurs extremely rarely. The flowers, white with delicate purple markings within, are very fragrant. The berries are bright red. The roots, leaves and berries contain powerful drugs similar to digitalis that affect the heartbeat and cause dizziness if they are eaten.

8a Leaves mostly basal or appearing so, or absent when flowering 9

8b Leaves alternate, spiral or whorled on the stem 15

9a Leaves 2, apparently basal, usually mottled; flowers solitary on a scape, nodding 10

9b Leaves more than 2, mostly basal but a few reduced up the stalk, not mottled; flowers several to many in an umbel or raceme 11

10a Flowers yellow; stigmas completely united (Fig. 95) .... YELLOW ADDER'S-TONGUE, *Erythronium americanum* Ker-Gawl.

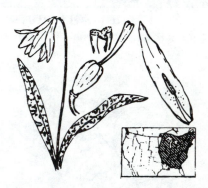

Figure 95

Damp woods. More common toward the East coast where it forms dense colonies of sterile single-leaved plants and flowering double-leaved ones. The two leaves are smooth and thick, lanceolate, 2 to 8 inches (5-20 cm) long and nearly stemless. The purplish-brown mottling on the leaves has suggested the name Fawn-lily. The corm is deep underground. The flower is 1 to 2 inches (2.5-5 cm) long.

10b Flowers white or pinkish white; stigma branches spreading (Fig. 96) .... **WHITE FAWN-LILY,** *Erythronium albidum* Nutt.

Figure 96

Wooded areas. Very closely resembles E. *americanum* except for the whitish flower and spreading stigmas. This species is more common in the western portion of its range.

Adder's Tongue, Trout Lily, Fawn Lily and Dog-tooth Violet are applied as names for these two plants. The first three are less confusing than the last, since these plants are lilies, not violets.

11a Flowers in an umbellate inflorescence (see Fig. 99) **12**

11b Flowers in a raceme (Fig. 97) **14**

Figure 97

12a Plants with onion-like odor, flowers pink or white, often intermixed with bulblets (Fig. 98) .... **MEADOW GARLIC,** *Allium canadense* L.

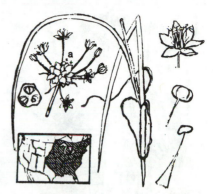

Figure 98

Moist meadows. Plant 10 to 20 inches (25-51 cm) high with narrow leaves and an erect umbel of pink or white flowers. Usually the umbel consists of many bulblets (a), which often replace all of the flowers.

There are several varieties of this species differing in the color of the flowers and the extent to which bulblets are formed.

Several species of this genus grow where cattle pasture and when eaten by them impart an undesirable flavor to the milk. Bulbs and bulblets may be eaten in the same manner as onions and garlic.

One interesting species in the eastern range, *Allium tricoccum* Ait., WILD LEEK or RAMPS, has lanceolate to elliptic leaves in early spring which die before the white flowers are produced. The species is also edible but very "aromatic."

The STRIPED GARLIC A. *cuthbertii* Small, has white flowers and no bulblets. It is not as large as the preceding; it grows from Florida to North Carolina.

12b Plants without onion-like odor and without bulblets in the inflorescence **13**

13a Leaves very narrow; umbel with 2 membranous bracts (Fig. 99) .... FALSE GARLIC, *Nothoscordum bivalve* (L.) Britt.

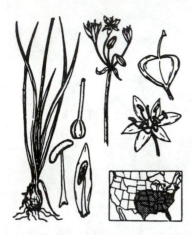

Figure 99

This plant bears 6 to 12 yellowish-green flowers about a half inch (1.3 cm) across in an umbel. The leaves are very narrow and usually shorter than the scape. The fruit is a somewhat flattened capsule.

This plant grows in sandy soil; in appearance it resembles the onions but does not have the onion odor.

There are two other species in this genus in North America and several more in South America.

13b Leaves oblong, elliptic or ovate; umbel without bracts (Fig. 100) .... YELLOW CLINTONIA, *Clintonia borealis* (Ait.) Raf.

Figure 100

Cold, moist woods. A handsome plant 6 to 15 inches (15-38 cm) high with yellowish flowers and 2 to 4 shiny leaves pointed at the tip. The 3 to 8 flowers in an umbel are nodding. The tepals are 1/2 to 3/4 inch (1.5-1.8 cm) long. The berries are pure blue—a color rare in nature. Another name is Corn Lily.

The WHITE CLINTONIA, *C. umbellulata* (Michx.) Morong, with 10 to 12 white flowers speckled with green or purplish dots, and only half as large as those of *C. borealis*, has black fruit. It grows in the eastern mountains and rich woods from New York and New Jersey west to Ohio and south to Georgia. Another name is Speckled Wood-lily.

Another species of this genus, *C. uniflora* (Menzies ex Schult.) Kunth, occurs from Washington to California and there is a species native to Japan.

14a Flowers 3 to 7, white inside, green with white margins outside; perianth segments 1/2 to 3/4 inches (1.3-1.9 cm) long (Fig. 101) .... STAR-OF-BETHLEHEM, *Ornithogalum umbellatum* L.

14b Flowers many (10 to 44), usually blue, if white then without green outer color; perianth segments usually less than 1/2 inch (1.3 cm) long (Fig. 102) .... WILD HYACINTH, *Camassia scilloides* (Raf.) Cory

Figure 101

Figure 102

Star-of-Bethlehem grows from a rather small membranous-coated bulb and attains a height of 12 inches (30 cm) or more. The several flowers measure from a half to almost an inch (1.3-2.5 cm) across. This European plant (a) is cultivated in our country and has frequently asserted its independence and gotten out on its own.

The DROOPING STAR-OF-BETHLE-HEM *O. nutans* L. (b) has larger flowers which are borne in a raceme and which do not stand so stiffly erect as with the above species. It too, is of European origin and sometimes escapes but is less common.

Scape 1 to 2 feet (30-61 cm) high, often bearing 1 or 2 narrow leaves; basal leaves up to 4 inches (10 cm) wide, shorter than the scape; flowers blue or whitish, the tepals with 3 to 7 parallel veins. It grows along streams and in open fields and woods. Previously this species was known as *Camassia esculenta* (Ker.) Robinson.

Scilla sibirica Andr., SQUILL, also has a blue-flowered raceme but the flowers are only 2 to 6, and the tepals have just one vein. It has escaped from cultivation in the Northeast and in Wisconsin and Missouri.

15a (8) Stem leaves whorled **16**

15b Stem leaves arranged alternately or spirally along the length of the stem **24**

16a Stem leaves in 2 whorls; flowers 3-9 in a terminal umbel, perianth segments essentially similar (Fig. 103) .... **INDIAN CUCUMBER-ROOT,** *Medeola virginiana* L.

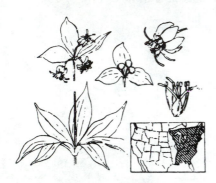

Figure 103

Rich damp woods. Entire plant 1 to 2 1/2 feet (30-76 cm) high. The leaves in the upper whorl usually 3, in the lower, 5 to 11. The greenish-yellow flowers on long pedicels mature into purple-black berries. The six stamens and the three recurved stigmas are reddish orange. The rhizome tastes like cucumber, hence the common name.

16b Stem leaves in one whorl of 3 leaves; flowers solitary, perianth of 3 green sepals and 3 colored petals (genus TRILLIUM) **17**

17a Flower sessile (without a stem) **18**

17b Flower peduncled (with a stem) **19**

18a Leaves petiolate; petals clawed (a) (Fig. 104) .... **PRAIRIE WAKE-ROBIN,** *Trillium recurvatum* Beck.

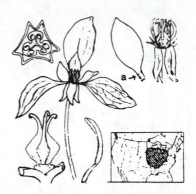

Figure 104

Wooded areas. A stout plant, 6 to 18 inches (15-46 cm) high, with dark reddish-purple petals. The petals are lanceolate and stand erect, while the sepals are recurved so that the tips point downward. The leaves are usually mottled with maroon and the ovary is 6-angled.

Most species of *Trillium* occasionally produce forms with flowers which are a different color from the usual, so that normally red-flowered species may actually have whitish flowers.

18b Leaves sessile; petals sessile (Fig. 105) .... **STEMLESS TRILLIUM,** *Trillium sessile* **L.**

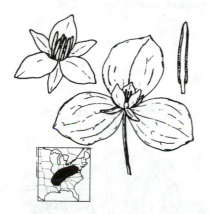

Figure 105

Moist woods. The genus name is derived from a word meaning triple, characteristic of all parts of the plant. This species is 5 to 10 inches (13-20 cm) high. The leaves are often mottled. The maroon-purple petals are narrow and, unlike the species above, sessile. The sepals in this species are erect or spreading. These fragrant flowers have also been given the less appealing common name Toadshade.

19a Leaves sessile or nearly so; ovary and fruit 6-angled, winged **20**

19b Leaves petiolate; ovary and fruit 3-lobed, not winged **22**

20a Petals usually purple-red; flower ill scented (Fig. 106) .... **ILL-SCENTED WAKE-ROBIN,** *Trillium erectum* **L.**

Figure 106

Rich woods. A very common eastern species 7-15 inches (18-38 cm) high. The rhombic, almost stemless leaves are abruptly pointed. Carrion flies are attracted by the spoiled-meat odor of the flower and thus assist with pollination. The flowers vary from dark purple to pink, greenish or white. The anthers are much longer than the filaments. The ovary is purple.

This species has many local common names referring to its color or smell. One, Birthwort, refers to the Indian use of tea from the rhizome of this (and other species) to aid childbirth.

20b Petals white, turning to pink or rose; flowers not ill scented **21**

21a Flower drooping beneath the leaves; petals recurved (Fig. 107) .... NODDING TRILLIUM, *Trillium cernuum* L.

21b Flower erect; petals spreading but not recurved (Fig. 108) .... LARGE-FLOWERING TRILLIUM, *Trillium grandiflorum* (Michx.) Salisb.

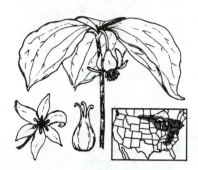

Figure 107

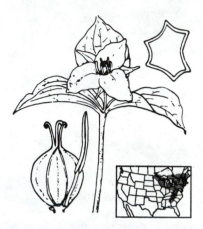

Figure 108

Moist woods. Often the flower is hidden beneath the broadly ovate leaves. Plant 8 to 14 inches (20-36 cm) high. Leaves net-veined as in all of the genus *Trillium*. The petals are white or pink and the ovary is white or pinkish. The anthers are pink to purple and the fruit is a dark red berry.

Rich woods. A tall stout plant 10 to 18 inches (25-46 cm) high with broad waxy-white petals 1 to 1 1/2 (2.5-3.8 cm) long, which later turn pink. Flowers appear later than the other trilliums. The fruit is black and slightly 6-lobed.

Trillium is another of those genera in which there are many species in the eastern woodlands, several species native in the mountains along the west coast, and more species native in Japan and eastern Asia. The climates of eastern North America and eastern Asia are similar enough that the same genus can grow in both places. It is no accident that many of our eastern ornamental shrubs are natives of Japan and China.

22a Flower nodding; petals rose colored (Fig. 109) .... ROSY TRILLIUM, *Trillium catesbaei* Ell.

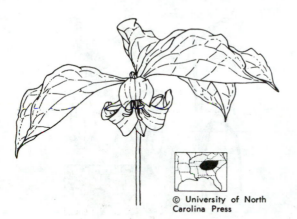

Figure 109

Moist wooded slopes. Another name for this plant is Bashful Trillium, alluding to the rosy blush of the petals and the way the flower hangs down and hides underneath the leaves. The leaves are ovate or lanceolate, narrowed into petioles at the base and with pointed tips, 2 3/8 to 6 inches (6-15 cm) long and 1 1/2 to 4 inches (4-10 cm) wide. The sepals, petals and anthers are recurved at maturity. This is a common trillium in the Southeast.

22b Flower erect; petals white 23

23a Plant 2 to 6 inches (5-15 cm) high; leaves and petals obtuse at the tips (Fig. 110) .... DWARF WHITE TRILLIUM, *Trillium nivale* Riddell

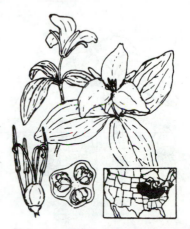

Figure 110

Rich woods. A dwarf trillium which often blooms while snow is yet on the ground and is sometimes called Snow Trillium. The ovary and fruit have 3 rounded angles instead of the 6 angles or wings of many other trilliums.

23b Plant 8 to 20 inches (20-51 cm) high; leaves and petals pointed; flower white with crimson stripes or a **V** on the petals (Fig. 111) .... **PAINTED TRILLIUM,** *Trillium undulatum* Willd.

25a Flowers small, many in a panicle; perianth segments very similar (Fig. 112) .... **MERRY-HEARTS,** *Zigadenus nuttallii* Gray

Figure 111

Figure 112

Cold damp woods. Very common in its northern range. One of the most beautiful trilliums, with a crimson V-shaped mark on the white petals. The leaves are ovate and taper to a point. The three-angled obtuse berry is a shining bright red when mature.

24a **(15) Leaves near base long and narrow, upper leaves becoming shorter** **25**

24b **Leaves elliptic to ovate, nearly the same size throughout the length of the stem** **26**

Long, narrow, grass-like leaves and sturdy scape bearing a sizeable panicle of greenish-white flowers, arise from a bulb-like base and measure 2 to 4 feet (61-122 cm) in height. The tepals have a short claw at the base and the lower third is covered by a gland. The several other species flower later in the season; all are somewhat poisonous. A common name for the members of the genus is Death Camas.

In the early spring this species is difficult to distinguish from the Wild Onion, which is edible. Pioneers and Indians were poisoned by mistakenly including Death Camas bulbs with the onions, an error more unfortunate because there is more of the poisonous alkaloid in the bulb than in the rest of the plant. Symptoms of the poisoning run from vomiting to diarrhea and decreased heartbeat.

25b Flowers large, few in an umbel; perianth segments unlike, sepals narrow, petals broad (Fig. 113) .... SEGO LILY, *Calochortus nuttallii* Torr. & Gray

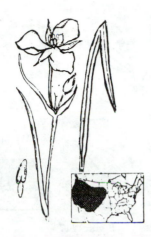

Figure 113

This lily may attain a height of 15 inches (38 cm). The flowers are in an umbel-like inflorescence but only one flower opens at a time. Each flower is white, lilac, or yellowish, with a purple spot at the base of each petal and has a diameter of 2 to 3 inches (5-7.5 cm). It is the state flower of Utah. The genus is well represented in the West with some showy members. The bulbs of this plant were eaten by the Mormons on their westward trek. The whole plant can be boiled and eaten, but it is better to appreciate this plant visually rather than gastronomically. This species and the previous one were named in honor of Thomas Nuttall, a plant explorer of the early American West.

26a Flowers many in a terminal panicle or raceme (Fig. 114) 27

Figure 114

26b Flowers solitary, paired, or in a 2- to 3-flowered umbel, terminal or axillary 29

27a Flowers yellowish green, borne in a panicle 8 to 20 inches (20-51 cm) long (Fig. 115) ....
.... AMERICAN WHITE HELLEBORE, *Veratrum viride* Ait.

Figure 115

The broad, oval, veiny leaves are arranged spirally on the stem. A blooming plant looks somewhat like a tasseled cornstalk. Before 1942 the drug from this plant had had long use in American prescriptions for hypertension. Its use was discontinued because there is a fine line between a therapeutic dose and poisoning. Leaves, roots and seeds contain alkaloids which cause digestive upset, major effects on heart and blood vessels, and death. A related species in California, *V. californicum* Durand has been called one of the most poisonous plants of the Sierras.

27b Flowers white **28**

28a Inflorescence a panicle; leaves broad or ovate (Fig. 116) .... **FALSE SOLOMON'S-SEAL,** *Smilacina racemosa* (L.) Desf.

Figure 116

Moist woods. The crowded panicle of small white fragrant flowers is the most attractive part of this common plant 1 to 3 feet (30-91 cm) high. The leaves are 3 to 6 inches (8-15 cm) in length and are coated beneath with fine pubescence. The fruit when ripe has purple specks on red and is aromatic.

28b Inflorescence a raceme; leaves lanceolate (Fig. 117) .... **STAR-FLOWERED SOLOMON'S-SEAL,** *Smilacina stellata* (L.) Desf.

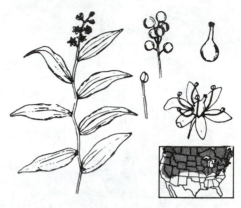

Figure 117

Moist banks. Also in Europe. The plant, 8 to 16 inches (20-40 cm) high, is smaller than *Smilacina racemosa,* but has larger flowers which are 3/8 inch (1 cm) wide, white, and are 8 to 10 in number. The berry is about 1/4 inch (0.6 cm) in diameter and is dark red when ripe. The leaves of this and the preceding species are borne in two lines opposite each other on the stems so that the whole plant has the appearance of a flat compound leaf.

THREE-LEAVED SOLOMON'S-SEAL, *Smilacina trifolia* (L.) Desf., is a smaller but similar plant of cold bogs in northeastern United States and eastern Canada. It has usually 3 (1 to 4) leaves and 3 to 8 flowers.

29a Anther linear, much longer than the filament; fruit a capsule **30**

29b Anther shorter than the filament, pointed or oblong; fruit a berry **32**

30a Leaves sessile; style much longer than the stamens, about the same length as the tepals (Fig. 118) .... **WILD OATS,** *Uvularia sessilifolia* **L.**

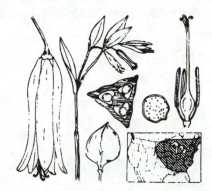

Figure 118

Sandy woods. This dainty plant, 6 to 12 inches (15-30 cm) high, bears lanceolate-oblong leaves. The 3/4 inch long (1.9 cm) flowers are greenish yellow and lightly fragrant. The capsule tapers at each end and is 3-winged.

The MOUNTAIN BELLWORT, *U. puberula* Michx., with leaves rounded at the base instead of pointed and styles separate for 1/2 their length, is found in the eastern mountains.

30b Leaves perfoliate; stamens slightly longer or shorter than the style; style 3/5 the length of the tepals or less **31**

31a Sepals and petals glandular-pubescent within; stamens shorter than the style (Fig. 119) .... **PERFOLIATE BELLWORT,** *Uvularia perfoliata* **L.**

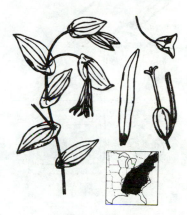

Figure 119

Wooded areas, usually acid soils. This bellwort, 5-16 inches (13-40 cm) high, has inch-long (2.5 cm), pale-yellow flowers, and glabrous leaves and stem. The flower has a delicate fragrance. The 3-parted capsule looks as if it had been abruptly cut off at the end.

This species flowers about 2 weeks later than *Uvularia grandiflora*.

31b Sepals and petals smooth within; stamens longer than the style (Fig. 120) .... **LARGE-FLOWERED BELLWORT,** *Uvularia grandiflora* Sm.

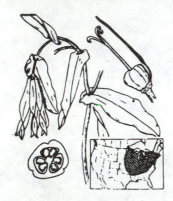

Figure 120

Rich woods. The plant is 6 to 18 inches (15-46 cm) high with drooping lemon-yellow flowers 1 to 1 3/4 inches (2.5-4.5 cm) long. The dark green leaves are finely pubescent beneath. The capsule does not end as abruptly as in *Uvularia perfoliata*.

32a Stem branched; flowers terminal, either solitary or 2-3 in an umbel; perianth segments at least 1/2 inch (1.3 cm) long; ovules few (Fig. 121) .... **HAIRY DISPORUM,** *Disporum lanuginosum* (Michx.) Nichols.

Figure 121

Rich woods. The plant, 1½ to 2½ feet (46-76 cm) high, has greenish-yellow flowers and red pulpy berries. Occasionally there are three flowers, but usually there are only two. The stamens are shorter than the perianth, and the 2- to 6-seeded fruit smooth. The leaves are hairy beneath. Another name is Yellow Mandarin.

The ROUGH-FRUITED DISPORUM, *Disporum trachycarpum* (S. Wats.) Benth. & Hook. f. with stamens as long as the perianth and with more seeds in the fruit ranges through South Dakota and Nebraska, and west to the Pacific.

32b Stem not branched; flowers axillary; perianth segments 3/8 inch (1 cm) long; ovules many **33**

33a Flowers rose purple; stigma 3-parted; anther 2-pointed (Fig. 122) ....
..SESSILE-LEAVED TWISTED-STALK, *Streptopus roseus* Michx.

33b Flowers greenish white; stigma entire; anther 1-pointed (Fig. 123) ....
.... CLASPING-LEAVED TWISTED-STALK,
Streptopus amplexifolius (L.) DC.

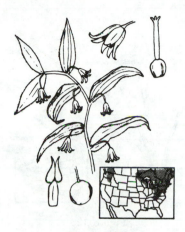

Figure 122

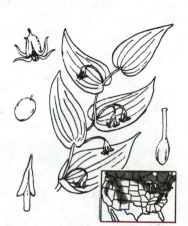

Figure 123

This plant grows from a rhizome to a height of 1 to 2½ feet (30-76 cm). The sessile leaves, 2 to 4½ inches (5-11 cm) long, are nearly clasping and fringed with hairs on the margins. The flowers are always single in the axil, about a half inch (1.3 cm) long, and are purple or rose. At maturity the berry is red and many seeded. The berries when eaten cause digestive upset, perhaps the reason for a common name Scoot-berry. The plant prefers moist woods and may be found well to the top of the Southeastern mountains.

Streptopus amplexifolius has a much wider range than *Streptopus roseus*. It grows from Greenland to Alaska and south as far as Pennsylvania and in the mountains to New Mexico and North Carolina. Plant 1 to 3 feet (30-91 cm) high. The leaves are glaucous beneath and the bases are heart shaped and clasping. The leaf margins are entire or with a few small teeth. The flowers are 1 to 2 on a peduncle, about a half inch (1.3 cm) long, greenish white. The berry is oval, a good half inch (1.3 cm) in diameter and with many seeds.

AMARYLLIDACEAE

AMARYLLIS FAMILY

1a Flowers several (2 to 6), yellow inside and green outside; plant villous (Fig. 124) .... **YELLOW STAR-GRASS,** *Hypoxis hirsuta* (L.) Coville

Figure 124

1b Flower solitary, pink or white; plant glabrous (Fig. 125) ....
.... **ATAMASCO LILY,** *Zephyranthes atamasco* (L.) Herbert

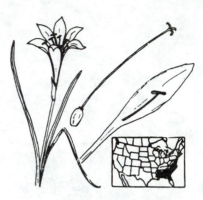

Figure 125

Meadows and open woods. The flower has spreading tepals that are a beautiful clear yellow inside, green and pubescent outside. The plant is perennial from a corm and produces basal grass-like leaves up to a foot (30 cm) long.

There are three other species of the genus in savannahs in the Southeast, all of which are rather rare. This species also has relatives in southern Africa and Australia.

Moist or sandy places, in woods or along roadsides. A very pretty plant 6 to 15 inches (15-38 cm) high but not widely distributed. The leaves are long and narrow and the 3-inch long (8 cm) flowers appear much like those of the true lily. The genus is placed in this family because of the inferior position of the ovary. The bulbs are poisonous when eaten in even small quantities.

There are three less common species in the southeastern United States and several more in the West and in Mexico and the Caribbean. Other names for these plants are Rain Lily or Zephyr Lily because they seem to bloom following rains.

In this family also are the Narcissus and Daffodils that make their show in spring gardens. These plants may be "naturalized" in woodsy settings and persist for years in abandoned gardens.

AGAVACEAE

CENTURY PLANT FAMILY
(Fig. 126) .... SOAPWEED,
Yucca glauca Nutt. ex Fraser

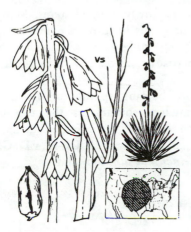

Figure 126

Other names are Beargrass and Spanish Bay-onet. Dry habitats. Leaves very stiff, 1 to 3 feet (30-91 cm) long, all basal and very sharp pointed, ½ inch (1.3 cm) wide, usually with many marginal threads. Flowers greenish white, drooping, 1 to 3 inches (2.5-15 cm) broad, in a narrow panicle or raceme up to 6 feet (1.8 m) tall; style green, short; capsule up to 3 inches (7.5 cm) long, 6-sided and erect.

A number of other Yucca species are native in the South and cultivated beyond their natural ranges. ADAM'S-NEEDLE, *Yucca filamentosa* L., with wider leaves and a more branched inflorescence, is the species most commonly cultivated and escaped in the eastern United States. Yuccas are especially abundant in the Southwest. The flowers and immature fruits can be eaten raw or boiled but they are somewhat bitter. Fiber products from the leaves include rope, baskets, and whisk brooms. The sharp point of the leaf with fibers attached makes a needle and thread. Crushed roots contain saponins which make a lather in water and can be used to wash laundry and as shampoo. Pollination is chiefly by the yucca moth, which stuffs pollen into the stigma of the pistil in which she lays her eggs, thus assuring seeds to feed the larvae.

IRIDACEAE

IRIS FAMILY

1a Perianth segments equal or nearly so; stigmas slender or thread-like; branches of style alternating with the anthers 2

1b Perianth of 3 outer spreading or reflexed segments (sepals) and 3 inner, usually smaller, erect segments (petals); style branches 3, petal-like, arching over the stamens (genus IRIS) 4

2a Flower about 2 inches (5 cm) broad; perianth segments with obtuse tips; filaments mostly free from each other (Fig. 127) .... PRAIRIE IRIS, *Nemastylis geminiflora* Nutt.

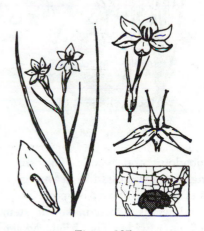

Figure 127

This interesting plant rises to a height of 1 to 2 feet (30-61 cm) from a scaly dark-colored bulb. The 3 to 4 leaves, usually 3/8 inch (1 cm) wide, are up to 10 inches (25 cm) long. The

1 to 2 light blue or purplish flowers about 2 inches (5 cm) across arise from a slender bract. The anthers and pollen are bright yellow. This flower can be found in bloom in late March and April; by June the fruits are reaching maturity.

2b Flower 1/2 to 3/4 inch (1.3-1.9 cm) broad; perianth segments with pointed tips; filaments fused their whole length 3

3a Flower-bearing spathes on long peduncles, 2 to 3 arising from a bract at the top of the scape (Fig. 128) .... **POINTED BLUE-EYED GRASS,** *Sisyrinchium angustifolium* P. Mill.

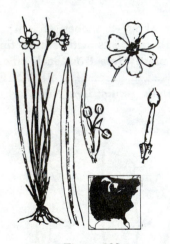

Figure 128

Moist meadows and woods. This stiff plant with bright green grass-like leaves grows to a height of a foot (30 cm) or more. The stems have broad wings which make the stem 1/4 inch (0.6 cm) wide. The bright violet-blue flowers a half inch (1.3 cm) across arise from a bract at the top of the peduncle and later develop into small spherical capsules.

These plants are somewhat misnamed. Although they appear grassy, they are not a grass, and although the flowers are often blue or violet, the eye in the center of the flower is yellow or yellowish green.

ATLANTIC BLUE-EYED GRASS, S. *atlanticum* Bickn., is similar and grows in the same range. It differs mainly by its light green color and narrowly winged stems. Many species of this genus have been named, but they so closely resemble each other that considerable expertise is needed to distinguish them.

3b Flower-bearing spathes sessile or short-peduncled, solitary at the top of the scape (Fig. 129) .... **BLUE-EYED GRASS,** *Sisyrinchium albidum* Raf.

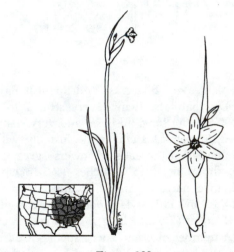

Figure 129

These pale green plants have a scape that is 3/16 inch (0.2 cm) wide, with narrow or broad wings. The flower is white to pale blue.

A closely similar species, S. *campestre* Bickn., may be locally more common. It has very narrow scapes and pale blue to white flowers about 1/2 inch (1.3 cm) across. Both of these species are found on sandy soils, in prairies or open woods.

(genus IRIS)

4a   Flowers reddish brown (Fig. 130) ....
.... COPPER IRIS,
Iris fulva Ker-Gawl.

Figure 130

The inflorescence rises to a height of 2 to 4 feet (61-122 cm) from a fleshy rhizome. Two to four narrow leaves about a half inch (1.3 cm) wide and several flowers, reddish brown with markings of blue and green, grow from each stalk. The flowers are glabrous and without a crest. It grows in swamps.

4b   **Flowers bluish, usually variegated with yellow and white** 5

5a   **Stems low, 2 to 6 inches (5-15 cm) high; sepals and petals of nearly equal length; 1- to 3-flowered** 6

5b   **Stems tall and leafy; sepals usually larger than the petals** 7

6a   Sepals crested; leaves lanceolate (Fig. 131) .... DWARF CRESTED IRIS, *Iris cristata* Soland.

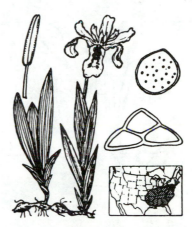

Figure 131

Rich woods. The short stem bears 1 or 2 flowers. The leaves are longer than the stem and up to an inch (2.5 cm) in width. The flowers are light violet to purple. Each one of the three sepals bears a 3-ridged yellow or white crest set in a field of white. The petals are upright or spreading.

Iris cristata should not be confused with the dwarf European iris, *Iris pumila* L., often cultivated in yards and gardens, which takes various shades from white, to pale blue to a rich royal purple. It, too, seldom exceeds 5 or 6 inches (13-15 cm) in height.

6b Sepals crestless, slightly hairy at base; leaves linear (Fig. 132) .... **DWARF IRIS,** *Iris verna* L.

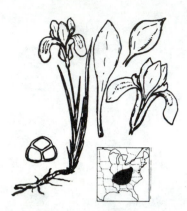

Figure 132

This plant, which may attain a height of 3 to 6 inches (8-15 cm), belongs to the wooded hillsides. The leaves are about a half inch (1.3 cm) wide; the flowers are violet-blue or sometimes white, with an orange or yellow streak bordered by purple on the sepals. Iris rhizomes contain a glycoside reported to poison livestock if eaten in quantity.

7a Ovary and capsule sharply 3-angled; leaves narrowly linear, 1/8 to 3/8 inch (0.3-1 cm) wide (Fig. 133) .... **SLENDER BLUE FLAG,** *Iris prismatica* Pursh

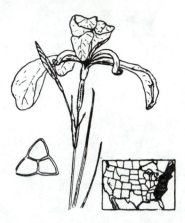

Figure 133

Marshes and wet places. This is a very slender-stemmed, narrow-leaved species, 1 to 3 feet (30-91 cm) high, bearing several violet-blue flowers with white spots veined with dark purple on the sepals. The flowers are usually borne in pairs. The narrow capsule is 1 to 1½ inches (2.5-3.8 cm) long.

The iris most commonly seen in yards and gardens is *Iris X germanica* L., a native of Europe, which has been developed into many widely varying colors.

7b Ovary and capsule bluntly 3-angled; leaves wider than 3/8 inch (1 cm) **8**

8a Sepal blades ovate, with or without a greenish spot at the base of the broad part of the blade (Fig. 134) .... **LARGE BLUE FLAG,** *Iris versicolor* **L.**

8b Sepal blades obovate, with a yellow, hairy spot below the tip of the style branch (Fig. 135) .... **SOUTHERN BLUE FLAG,** *Iris virginica* **L.**

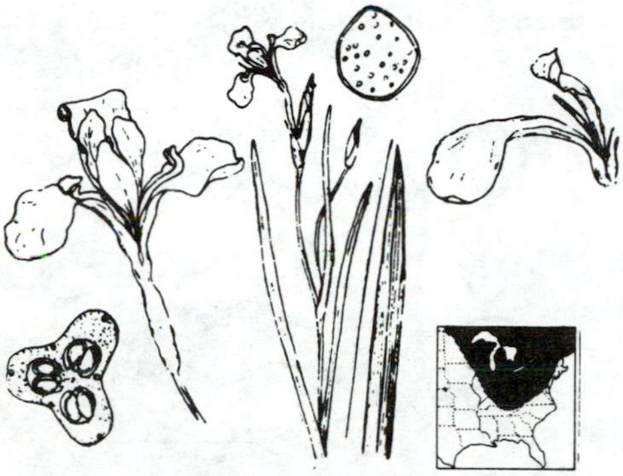

Figure 134

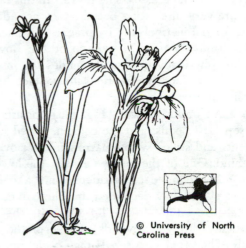

Figure 135

© University of North Carolina Press

Wet places. A common iris 1 to 3 feet (30-91 cm) high with smooth erect leaves and several flowers. The violet-blue perianth consists of 3 reflexed, spatulate sepals 2 to 3 inches (5-8 cm) long on the outside, and 3 erect smaller petals on the inside. The 3 divisions of the style are petal-like and arch over the 3 stamens. The 3-lobed capsule is 1 to 1½ inches (2.5-3.8 cm) long. Rhizomes of this species have been used medicinally as a purge or poultice. An overdose causes severe gastrointestinal upset.

These irises of swamps, ditches and other wet places are light violet to lavender with darker violet veins in the sepals and petals. The sepals are 3 inches (7.5 cm) long, spreading or somewhat recurved; the petals are erect or spreading and 2 inches (5 cm) long.

Another member of this family is the Crocus which blooms in early spring gardens. A later-blooming representative is the Gladiolus.

ORCHIDACEAE

ORCHID FAMILY

The Orchid family has more species than any other family of plants, with the possible exception of the grasses. The flowers in this family are very interesting because they are adapted for pollination only by precise movements of various insects. The shape of the flower either physically or visually guides the insect through the proper motions to contact both pollen and stigma.

An orchid flower (Fig. 136) consists of 3 sepals, 3 petals, and the column (fused filaments and style) with a narrow inferior ovary behind. One of the petals (usually the lowest one) is expanded or broadened into a sometimes elaborately decorated lip. The two other petals are lateral to the lip. Opposite the lip (usually directly upwards) is one of the sepals. The other two sepals may be lateral, or may be fused together under the lip.

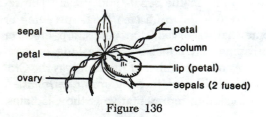

Figure 136

1a   Fertile anthers 2, located on either side of the column; lip a conspicuous inflated pouch (genus CYPRIPEDIUM) 2

1b   Fertile anther one, located at the tip of the column 6

**2a   Leaves basal, usually 2 (Fig. 137) ....
.... PINK LADY'S-SLIPPER,
Cypripedium acaule Ait.**

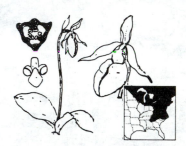

Figure 137

In acid soil in both swamps and dry, rocky woods. Also called Stemless Lady's-slipper and Moccasin Flower. The two leaves are elliptic and slightly hairy. The sepals and lateral petals are yellowish green or brownish. The delicate pink, drooping lip with its deeper red veins is 2¼ inches (5-6 cm) long, the longest of all our lady's-slippers. Plant 6 to 18 inches (15-46 cm) high. Many of the lady's-slippers grow in clumps, but this one prefers to scatter out.

2b   Leaves several, on the stem 3

3a The 3 sepals separate; lip whitish with crimson veins, the tip prolonged into a blunt spur-like "chin" below (Fig. 138) .... RAM'S HEAD LADY'S-SLIPPER, *Cypripedium arietinum* R. Br.

4a Sepals and lateral petals white, obtuse; the petals shorter than the lip (Fig. 139) .... SHOWY LADY'S-SLIPPER, *Cypripedium reginae* Walt.

Figure 139

Figure 138

This orchid grows to 12 inches (30 cm) high and bears only one flower. The three to five leaves are 2 to 4 inches (5-10 cm) long and about half as wide. The petals are greenish brown and very narrow. They and the sepals are both longer than the lip, which is red and white and which measures a little over a half inch (1.3 cm) in length. A spur-like part at the apex of the lip gives it the "ram's head" look.

Wet places, bogs or woods. These are large plants, often nearly 3 feet (90 cm) tall with hairy stem and leaves. The showy flowers are up to 3 inches (8 cm) across. The lip is basically white but usually heavily blushed with pink or rose, with darker purple spots inside. The sepals are white, oval, and obtuse, the lower two united. The lateral petals are also white, but narrower than the sepals. This species begins to bloom later than the others.

3b The two lower sepals at least partly united and held below the lip; lip not prolonged at the tip 4

4b Sepals and lateral petals yellowish green and streaked with brown or purple, or brownish, pointed; the petals longer than the lip and twisted 5

5a Lip yellow (Fig. 140) ....
.... YELLOW LADY'S-SLIPPER,
Cypripedium pubescens Willd.

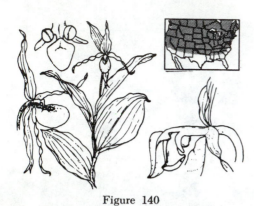

Figure 140

5b Lip white (Fig. 141) ....
.... WHITE LADY'S-SLIPPER,
Cypripedium candidum Muhl. ex Willd.

Figure 141

Rich woods and bog-lands. The inflated, golden-yellow lip 1½ to 2 inches (3.8-5 cm) long is the most attractive feature of the plant, which is 9 to 27 inches (23-69 cm) high. Two of the three petals are spirally-twisted and lateral; the third one forms the lip. The sepals are three; an upper one and two that are united into one below the lip with separate tips. This is the most common of the lady's-slippers.

This species has also been referred to as a variety (var. *pubescens*) of *C. calceolus,* a closely related species which occurs in Europe and Asia. In the early United States Pharmacopoeia, *Extractum Cypripedi, Fluidum,* a root extract, was listed as a sedative. Two of its many common names are Whippoorwill-shoe and Golden-slipper.

This beautiful little lady's-slipper grows to a height of 6 to 12 inches (15-30 cm) and is found in bogs and meadows. It produces 3 or 4 leaves about 1½ inch (3-8 cm) wide and up to 5 inches (13 cm) long. The purple spotted green sepals and the greenish petals are only a trifle longer than the lip (slipper). The lip is almost an inch (2.5 cm) long, waxy white with purple stripes inside.

Should you travel to the mountains of the North American west coast as far north as Alaska, and into the mountains of Korea, Japan and China, you will encounter other, and quite interesting species of this same genus.

6a (1) Leafless purplish-yellow saprophytes without a trace of green 7

6b Leaves one or more, sometimes absent at flowering time; stem green 9

7a Stem yellowish; lip 3-lobed (the lateral lobes smaller), usually neither spotted nor striped (Fig. 142) .... EARLY CORAL-ROOT, *Corallorhiza trifida* (L.) Chatelain

Figure 142

This orchid is found in wet woods of cedar, poplar, or spruce and must grow where it can find sufficient organic material in the soil. The yellow or greenish-yellow stem is 4 to 10 inches (10-25 cm) high and has 2 to 5 close-fitting sheath-like scales which are the remnants of the once functional leaves. The flowers are yellowish white to dull purple. The petals and the top sepal bend together and form a hood over the column. The much branched rhizome absorbs the food from the leaf mold. The capsules (a) hang downward when mature.

7b Stem tan to reddish purple; margin of lip wavy or finely toothed, but not 3-lobed; flowers either spotted or striped **8**

8a Lip white with reddish spots; margin of the lip toothed (Fig. 143) .... WISTER'S CORAL-ROOT, *Corallorhiza wisteriana* Conrad

Figure 143

The flowering stem grows to a height of 15 inches (38 cm) and bears 6 to 15 flowers in a raceme at its top. Sepals and petals yellowish, tinged with maroon. The white lip is about a third of an inch (0.8 cm) long with wavy margin, notch at end and speckled with red. This species blooms from February to July, south to north in its range. Another name is Spring Coral-root.

From June to October one might find blooming the similar LATE SOUTHERN CORAL-ROOT, *C. odontorhiza* (Willd.) Nutt. In that species the sepals have one vein instead of 3 as in *C. wisteriana.*

8b Lip white with purple stripes, the margins rolled inward (Fig. 144) ....
.... **STRIPED CORAL-ROOT,**
Corallorhiza striata **Lindl.**

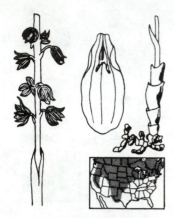

Figure 144

Moist soil and leaf mold. Flowering stem 10 to 20 inches (25-51 cm) bearing a raceme of 10 to 25 flowers. The sepals and petals are pinkish yellow or white, striped with dark purple. The capsules are nearly an inch long. The coral-shaped rhizome accounts for the generic and common names.

9a Flowers solitary (rarely 3), showy, pink, rose-purple, or purple and yellow **10**

9b Flowers several in a spike or raceme, yellowish green, greenish white, or purple and white **13**

10a Leaves 5 or 6 in a whorl near top of plant; flower purple and greenish yellow (Fig. 145) .... **WHORLED POGONIA,**
Isotria verticillata (Muhl. ex Willd.) **Raf.**

Figure 145

This species grows in moist woods and has a stem which reaches about a foot (30 cm) high, bearing leaves 1 to 3 inches (2.5-8 cm) long in a whorl of 5 or 6. The flower has a 3-lobed yellow and purple lip. The sepals and petals are purplish brown, narrow, and about twice as long as the yellowish green, elliptic petals. The inch-long (2.5 cm) capsule stands erect.

10b Leaves 1 to 3, basal or on the stem, appearing with or after the flower **11**

11a Lip inflated, slipper-like with an overlap forming an apron in front (a), 2-cleft at tip (Fig. 146) .... CALYPSO, *Calypso bulbosa* (L.) Oakes

Figure 146

Deep mossy woods. This lovely little orchid grows at most 7 inches (18 cm) high. Some call it Fairy-slipper. For its height, it has a large flower. The slipper part is 3/4 inch (1.9 cm) long. Sepals and lateral petals are pale purple, the lip is white to pale maroon with reddish spots and lines inside, and the apron is white with purple spots. The plant produces a single leaf in the fall which overwinters and dies after the flower has bloomed.

11b Lip crested, not inflated **12**

12a Leaves 2 or 3, lanceolate-oval, one leaf always midway on the stem (Fig. 147) ....
.... ROSE POGONIA, *Pogonia ophioglossoides* (L.) Juss.

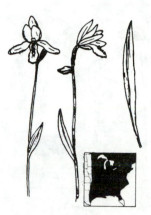

Figure 147

Sphagnum bogs and swamps. This orchid 8 to 15 inches (20-38 cm) high, with its daintiness of form, delicate fragrance and rose color, is a favorite. The lip is blunt spoon shaped, narrowed toward the base. The expanded portion of the lip is pink, deeply fringed on the edges and centered with a brush of yellow hairs. The other petals and the sepals are rose-pink to white. In northern bogs this species may be found by the hundreds.

12b Leaf solitary, basal, grass-like, often appearing after the flower (Fig. 148) .... ARETHUSA, *Arethusa bulbosa* L.

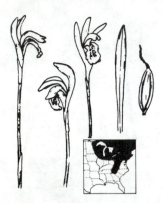

Figure 148

Sphagnum bogs. A beautiful orchid, 5 to 12 inches (12-30 cm) high, similar to Rose Pogonia in habitat and appearance. The lip of Arethusa also has its tuft of hairs, but is bent downward in a way to suggest the name Dragon-mouth. The tongue-like lip is pinkish, and striped and spotted with purple and yellow. The magenta sepals and other petals are erect or arching. It has been suggested that if the lip is the dragon's tongue, then two of these must be its ears. The solitary, showy flower is 1 to 2 inches (2.5-5 cm) high. Both *Pogonia* and *Arethusa* have counterparts in eastern Asia.

13a Flowers with a well developed, distinct, long or short spur 14

13b Flowers spurless 16

14a Leaves on the stem; flowers greenish; spur short, scale-like (Fig. 149) .... LONG-BRACTED ORCHIS, *Coeloglossum viride* Hartman var. *virescens* (Muhl. ex Willd.) Luer

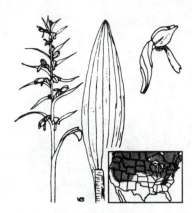

Figure 149

Synonym: *Habenaria viridis* (L.) R. Br. var. *bracteata* (Muhl. ex Willd.) A. Gray

In woods and open fields, this sturdy plant grows to a height of 6 inches to 2 feet (15-61 cm). The lower leaves are broadly oval and become slimmer and lanceolate higher on the stem. The bracts are longer than the flowers which arise in their axils, especially the lower ones. The flowers are green or greenish and about one-third inch (0.8 cm) long. The species is found in Europe as well as being rather widely distributed in North America.

14b Leaves basal; flower lavender and white; spur elongate 15

15a Leaves 2; lip white, not lobed or spotted (Fig. 150) .... SHOWY ORCHIS, *Galearis spectabilis* (L.) Raf.

Figure 150

Synonym: *Orchis spectabilis* L.

Rich woods. The plant is 4 to 10 inches (10-25 cm) high with 3 to 8 inch-long (2.5 cm) flowers. The over-arching hood is mauve; the 3/4 inch (1.9 cm) lip and 2/3 inch (1.7 cm) spur are white. The combination of mauve and white is a rarity among orchids. This species is fairly common for an orchid.

15b Leaf 1; lip white, lobed and spotted with purple (Fig. 151) .... ROUND-LEAVED ORCHIS, *Amerorchis rotundifolia* (Banks ex Pursh) Hultén

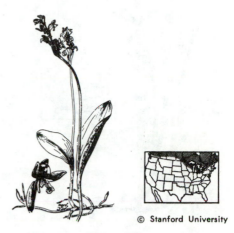

© Stanford University

Figure 151

Synonym: *Orchis rotundifolia* Banks ex Pursh

This plant of wet woods and evergreen swamps has a northern distribution, rare in the East, but more common westward. It has a single rounded leaf to 4 inches (10 cm) long and a raceme of 1 to several flowers. The white to pale purple petals and the top sepal form a hood over the column. The 3-lobed lip is white and spotted with purple, the middle lobe often notched.

16a Flowers close together in a single row, the row usually spiralled up the inflorescence; sepals and petals white, held close together so flower appears tubular 17

16b Flowers spreading, inflorescence not an evident spiral; sepals and petals spreading, green, purple, or yellowish 18

17a Leaves linear, present at flowering time, to 12 inches (30 cm) long; upper part of stalk, bracts and flowers hairy (Fig. 152) .... **SPRING LADIES'-TRESSES,** *Spiranthes vernalis* Engelm. & Gray

17b Leaves ovate to lanceolate, 2 1/2 inches (6.3 cm) long, usually absent at flowering; stalk and bracts not hairy (Fig. 153) .... **SLENDER LADIES'-TRESSES,** *Spiranthes lacera* Raf. var. *gracilis* (Bigelow) Luer

© Stanford University

Figure 152

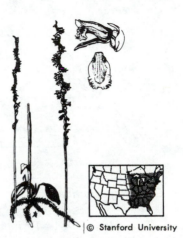

© Stanford University

Figure 153

Moist ground. These tall plants, up to 3½ feet (107 cm) high, have long, grass-like leaves 3/8 inch (1 cm) wide and a spiral of small white or yellowish flowers. The upper parts of the stalk and the flowers are hairy. This species grows in bogs, salt marshes, dunes, savannahs and in open woods. Its range extends as far south as Guatemala.

Sandy soil, dry woods and meadows. The leaves of this slender 2½ foot (76 cm) tall plant are basal and ovate or lanceolate, and up to 2½ inches (6.3 cm) long, but they are usually not present when the plant flowers. The small white flowers have a white lip with a green stripe in the center. This variety flowers from spring into fall.

18a Leaf solitary, large, usually absent at flowering or present as withered fragments; lip 3-lobed, white with maroon spots (Fig. 154) .... **PUTTYROOT,** *Aplectrum hyemale* (Muhl. ex Willd.) Nutt.

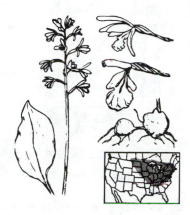

Figure 154

Rich deciduous woods. The 8-15 purplish-yellow flowers are on a scape 10 to 20 inches (25-51 cm) high. The quaint name of Adam-and-Eve is given because of the pair of corms fastened together horizontally. The leaf, about 7 inches (18 cm) long and up to 3 inches (8 cm) wide, is produced in late summer, and winters over. Usually it is withered by the time the flowers open.

18b Leaves 2, present at flowering; lip 2-lobed or not lobed but with a projection at the tip **19**

19a Leaves midway on the stem, opposite; lip deeply 2-lobed; lateral petals elliptic (Fig. 155) .... **HEART-LEAVED TWAYBLADE,** *Listera cordata* (L.) R. Br.

Figure 155

Bogs and damp mountainous woods. Plant 4 to 8 inches (10-20 cm) high with purplish or yellowish flowers. Not a common nor conspicuous orchid, but widely distributed. All of the genus *Listera* have a pair of leaves, "twayblades," situated midway on the stem and opposite each other. In this species the leaves are ovate-cordate. Also in Greenland, Iceland, Europe, and Japan.

19b Leaves basal; lip with a small point at the tip; lateral petals linear **20**

20a Flowers about 1/2 inch (1.3 cm) long; petals and lip reddish purple (Fig. 156) .... LILY-LEAVED TWAYBLADE, *Liparis lilifolia* (L.) L.C. Rich. ex Lindl.

20b Flowers about 1/4 inch (0.6 cm) long; petals and lip yellowish green (Fig. 157) .... LOESEL'S TWAYBLADE, *Liparis loeselii* (L.) L. C. Rich.

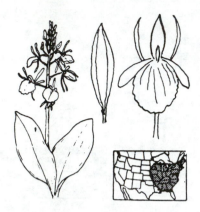

Figure 156

Figure 157

Mossy areas in moist woods. The plant may stand up to 10 inches (25 cm) high with broadly ovate leaves 3 to 7 inches (7.5-18 cm) long. The translucent flowers are 5 or more in a loose raceme. The sepals are greenish. The green-tinged lip is broadly wedge-shaped with a point at the apex.

This species is somewhat similar to the preceding. The flowers are about ¼ inch (0.6 cm) across and yellow green. The lip is narrow, with a small point at the tip. The capsule becomes almost ½ inch (1.3 cm) long. It grows in wet, cool places and is native of Europe as well as of this hemisphere.

SANTALACEAE

SANDALWOOD FAMILY

(Fig. 158) .... BASTARD TOAD-FLAX, *Comandra umbellata* (L.) Nutt.

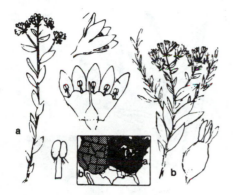

Figure 158

Comandra umbellata is a green parasite on the roots of plants growing in dry soil in most of the eastern half of the United States. The plant is 6 to 15 inches (15-38 cm) tall. The flowers have 5 greenish-white sepals. They may be perfect or only staminate.

PALE COMANDRA, *C. umbellata* ssp. *pallida* (A.DC.) Piehl (Fig. 158b) grows on roots of shrubs or herbs in dry soil in the western half of the United States. This subspecies is more glaucous than the eastern one and has less obvious lateral veins on the leaves.

ARISTOLOCHIACEAE

BIRTHWORT FAMILY

(Fig. 159) .... WILD GINGER, *Asarum canadense* L.

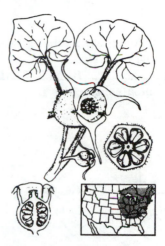

Figure 159

Woods. This low, woodland plant with heart-shaped or kidney-shaped leaves has purplish-brown flowers with a cup-like calyx 1 to 1½ inches (2.5-3.8 cm) broad, an inferior ovary and 12 stamens. The flower blooms near the ground, half hidden by the leaves. In the typical form the 3 calyx lobes are spreading, elongate, and gradually taper to a narrow tip. Variety *acuminatum* Ashe has longer lobes and var. *reflexum* (Bickn.) Robins. has triangular lobes which are folded back onto the calyx.

The underground stem has been used as a substitute for the ginger of commerce, which is a tropical monocot. Tea or other extracts from the rhizome have been used in early remedies to treat heart disease, ear infections, and coughs. The United States Pharmacopoeia included it among the "official" drug preparations from 1820-1870.

POLYGONACEAE

BUCKWHEAT FAMILY

1a   Sepals 4 to 5; stigmas 2 or 3 (genus POLYGONUM 2

1b   Sepals 6 in 2 rings of 3; stigmas 3 (genus RUMEX) 4

2a   Plant twining; calyx lobes greenish white (Fig. 160) ....
.... BLACK BINDWEED, *Polygonum convolvulus* L.

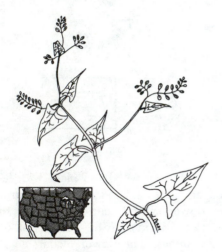

Figure 160

Roadsides and weedy places. This trailing plant has sagittate to nearly triangular leaves 3/4 to 2 3/8 inches (2-6 cm) long. The calyx is green outside and white inside. The flowers occur in clusters of 3 to 6 in straggly racemes. Also called Wild Buckwheat.

    The true Buckwheat, *Fagopyrum esculentum* Moench, is also a member of this family. Its dry fruit is treated as a cereal grain, but other cereals are members of the Grass Family.

2b   **Plant erect, or if prostrate not twining or trailing; calyx pink, rose or white 3**

3a   Sheaths around stem at leaf bases have long bristles on the margin (a); leaves often with a dark blotch in the middle; stigmas 2 or 3 (Fig. 161) ....
.... LADY'S-THUMB, *Polygonum persicaria* L.

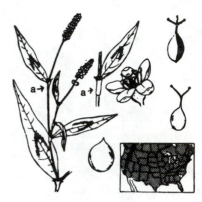

Figure 161

Damp waste areas. The compact, "thumb-like" racemes of small pink or rose flowers are terminal or axillary on plants up to 32 inches (81 cm) tall. The leaves are lanceolate and have hairs along the midrib. This very common plant has been introduced from Europe and is now well established throughout North America except in the extreme North. When eaten, this plant is reportedly peppery enough to make eyes water. Animals which graze large quantities may experience gastric disturbances but eventually recover.

3b Sheaths around stem without stiff marginal bristles; stigmas 2 (Fig. 162) ....
.... SMARTWEED,
Polygonum pensylvanicum (L.) Small

Figure 162

Disturbed, moist soil. This is a native, widespread weed that can grow as tall as 6 feet (1.8 m). The leaves are lanceolate as in the previous species but there are no bristles on the sheath. The flowers are pink to white.

A prostrate species of this genus, *Polygonum aviculare* L., KNOTWEED, also blooms along roadsides and in trampled lawns in spring. It is much branched and spreading. The greenish flowers with white or pink margins to the calyx are small and nearly hidden by the leaf sheath.

4a Leaves hastate; plant usually less than 10 inches high (25 cm); leaves with sour taste (Fig. 163) ...... SHEEP SORREL, *Rumex acetosella* L.

Figure 163

This weed which grows in large patches on acid soil and crowds out about everything else was introduced from Europe and is widely scattered. It grows to a height of 4 to 10 inches (10-25 cm); the flowers, often deep red, sometimes show much yellow. The vegetation has a sour taste.

Poisonous, edible, and medicinal characteristics have all been reported for this species. Livestock may eat too much and become ill, but humans, who value the leaves for use in salads or cooked as spinach or in soup, apparently do not consume toxic quantities. The sour tasting leaves mixed with water are employed in a lemonade-like drink and medicinally as an astringent.

4b Leaves lanceolate to ovate-lanceolate; plant 1 to 4 feet (30-122 cm) high 5

5a Leaves dark green, wavy margined; fruit rust colored with 3 papery wings (Fig. 164) .... CURLED DOCK, *Rumex crispus* L.

5b Leaves pale green, flat; fruit tan with 3 large corky wings. (Fig. 165) .... PEACH-LEAVED DOCK, *Rumex altissimus* Wood.

Figure 164

Figure 165

This is still another European weed, easily recognized by its lanceolate curly leaves which are 6 to 10 inches (15-25 cm) long and up to 2 3/8 inches (6 cm) wide. The normally green leaves often show reddish touches. The flowers are green and would not be noticed except that the inflorescence is large, up to 16 inches (40 cm) long, much branched, and with linear leaves mixed with the flowers. In fruit, the 3 inner russet sepals are large and winged, and surround the achene.

Large quantities of these leaves poison livestock. Young leaves can be eaten raw or steamed; old leaves must be boiled thoroughly in several changes of water. Seeds can be used like buckwheat. The juice is supposed to relieve itching. The root has laxative powers and was recognized in the United States Pharmacopoeia until 1890.

This native weed grows to 4 feet (122 cm). The pale green leaves are lanceolate and pointed. The green flowers are in an open panicle about 12 inches long.

Two other Docks, PATIENCE DOCK, *R. patientia* L. and BROADLEAF DOCK, *R. obtusifolius* L., also bloom in the spring in parts of the Midwest. Both species have broad leaves, the larger ones heart shaped at the base.

Rhubarb, *Rheum rhaponticum* L., is also a broadleaved member of the family. Its leaf blades are poisonous but the petioles make a tangy pie or cobbler.

NYCTAGINACEAE

FOUR O'CLOCK FAMILY

(Fig. 166) .... **WILD FOUR O'CLOCK,**
Mirabilis nyctaginea (Michx.) MacM.

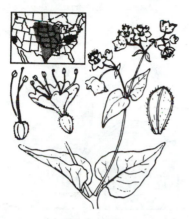

Figure 166

This stout Wild Four-o'clock is native in dry soil in the Middle West and West and is also found in the eastern part of the United States where it has been introduced from its native habitat. In this 1 to 5 feet (0.3-1.5 m) tall species, the flowers have no petals but the calyx of united sepals is pink or reddish purple and looks much like a corolla. The calyx is 3/8 inch (1 cm) long and is constricted above the ovary so that the ovary appears inferior. The ovary is superior. Several flowers are subtended by an involucre of united bracts. The involucre also has touches of pink and becomes enlarged after the flowers have opened. The leaves are ovate or cordate with petioles up to an inch (2.5 cm) long. Another common name is Umbrella-wort. Sometimes the species has been placed in the genus *Oxybaphus*.

AIZOACEAE

CARPET WEED FAMILY

(Fig. 167) .... **CARPETWEED,**
Mollugo verticillata L.

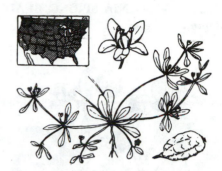

Figure 167

This inconspicuous but widely-spread plant is a native of the warmer parts of America but is now distributed as a weed over most of the United States and southern Canada. It grows prostrate on the ground in sandy river banks, roadsides, and waste places, sometimes forming patches 15 to 20 inches in diameter. The tiny whitish-green flowers are about 1/12 inch (0.2 cm) broad, with 3 or 4 stamens. The leaves are oblanceolate, 3/8 to 1 1/8 inch (1-3 cm) long, and in a whorled arrangement on the stem.

PORTULACACEAE

PURSLANE FAMILY

1a Stem leaves linear-lanceolate, tapered at the base so that a distinct petiole is hardly evident (Fig. 168) .... **VIRGINIA SPRING BEAUTY,** *Claytonia virginica* L.

Figure 168

This very common and early spring flower grows in moist open woods. The plant, 6 to 12 inches (15-30 cm) high, is often somewhat decumbent. The flowers, 1/2 to 3/4 inch (1.3-1.9 cm) broad, are occasionally white but are usually pink with dark pink veins. The style has 3 short branches. As in all of *Claytonia*, the flowers are in a raceme. Underground there is a corm. The bell-shaped calyx of 2 sepals becomes enlarged and turns red in fruit and is sometimes mistaken for a flower. The leaves are more than 3 inches (7.5 cm) long and 3/4 inch (1.9 cm) wide or less. Small bees visit these flowers.

1b Stem leaves ovate-lanceolate or ovate, with a distinct petiole (Fig. 169) .... **CAROLINA SPRING BEAUTY,** *Claytonia caroliniana* Michx.

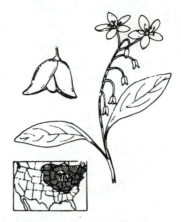

Figure 169

Moist woods. This species closely resembles *Claytonia virginica* except that it has broader leaves and somewhat smaller flowers. The plant is 5 to 10 inches (13-25 cm) high and has a raceme of pink flowers about 1/2 inch (1.3 cm) broad. The leaves are up to 2 3/4 inches (7 cm) long and 1¼ inch (3.2 cm) wide, with a distinct petiole.

In both of these species the corm is starchy and can be eaten either boiled or roasted. However the corms are so small that a good many plants must be destroyed to harvest even one moderate serving.

In the western states and British Columbia the WESTERN SPRING BEAUTY, *C. lanceolata* Pursh, also has pink flowers and ovate-lanceolate leaves, but the stem leaves have no evident petiole.

CARYOPHYLLACEAE

PINK FAMILY

"Pinks" are the carnations and their relatives and although many of them are pink or rose colored, other members of this family have white flowers.

1a Sepals united into a tube at least at the base 2

1b Sepals distinct or nearly so 8

2a Calyx lobes longer than the tube; petals purple-red, without a deep apical notch (Fig. 170) .... CORN COCKLE, *Agrostemma githago* L.

Figure 170

Disturbed soil, in grain fields and weedy places. This European native with narrow, hairy leaves grows to 3 feet (91 cm) tall. The calyx tube has 10 ribs and bears 5 long and narrow lobes which spread under the petals. The flowers, with 5 separate, purple petals, are up to 2½ inches (6.5 cm) broad.

2b Calyx lobes shorter than the tube; petals white, pink, or red; if red, with a deep apical notch (genus SILENE) 3

3a Flowers white or pink 4

3b Flowers red (Fig. 171) .... FIRE PINK, *Silene virginica* L.

Figure 171

Dry woods. This viscid-pubescent perennial attains a height of 1 to 2 feet (30-61 cm) and when flowering is a very showy plant. The basal leaves are usually spatulate and 4 or 5 inches (10-13 cm) long; the flowers are 1 to 1½ inches (2.5-3.8 cm) broad. The crimson petals have a pair of projections where the blade joins the claw, creating a crown-like ring at the center of the flower. The tips of the petals are 2-cleft, 2-lobed or irregularly cut. It is sometimes called Indian Pink.

4a Petals without a deep notch at the tip; plants 3 1/8 to 8 inches (8-20 cm) high; leaves mostly basal (Fig. 172) ....
.... **WILD PINK,**
Silene caroliniana Walt.

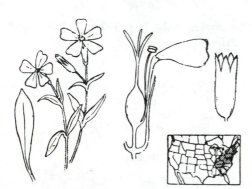

Figure 172

Dry sandy soil and rocky places. The plant grows 3 1/8 to 8 inches (8-20 cm) high with basal spatulate leaves and opposite lanceolate stem leaves. The flowers, about 1 inch (2.5 cm) broad, are in dense terminal cymes and have pink or white wedge-shaped petals, either entire at the tip or slightly notched or scalloped. The plant is viscid pubescent above and nearly glabrous below.

Two subspecies in addition to the typical subspecies have been named. Subsp. *pensylvanica* (Michx.) Clausen has a very glandular calyx and the claws of the petals are longer than the calyx. Subsp. *wherryi* (Small) Clausen has no glands on the calyx.

4b Petals with a deep notch at the tip; plants 12 inches (30 cm) or more high; stems leafy **5**

5a Calyx inflated, papery; plant glabrous (Fig. 173) .... **BLADDER CAMPION,**
Silene vulgaris (Moench) Garcke.

Figure 173

Synonym: *Silene cucubalus* Wibel.

Roadsides and weedy places. Introduced from Europe, this plant flowers in late spring. The stem grows to 3 feet (91 cm) tall and is clasped by lanceolate or oblanceolate leaves. The smooth, balloon-like calyx is pale green with obvious veins; the notched petals are white.

The young shoots are edible, if necessary, in the spring but the internal saponins (soapy substances) give the greens a bitter taste.

5b Calyx tubular or fusiform; plant hairy or glandular, at least at the nodes **6**

6a Styles **5** or occasionally **4** (flowers with either stamens or pistils); petals white (Fig. 174) .... **WHITE CAMPION,** *Silene pratensis* (Raf.) Godron

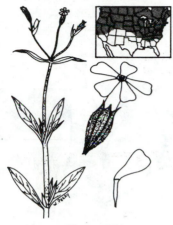

Figure 174

Synonym: *Lychnis alba* Mill.

Roadsides and fields. There are male and female plants of this 20 to 47 inch (50-120 cm) tall European species. The male plants have flowers with stamens only and 10 veins in the calyx; the female flowers have 5 or 4 styles and 20 veins in the calyx. In both, the petals are white and notched at the tip. The fragrant flowers open late in the day and close the next morning.

6b Styles **3,** or rarely **4** (flowers of some species either staminate or pistillate); petals white or pinkish **7**

7a Stem glandular-hairy all over; crown at center of flower obvious; flowers 1 inch (2.5 cm) across (Fig. 175) ....
.... **NIGHT-FLOWERING CATCHFLY,** *Silene noctiflora* L.

Figure 175

Sandy soil. This is also a night-flowering species, introduced from Europe, and very similar to *Silene pratensis.* The chief difference is that there are 3, or sometimes 4, styles in this species and five in S. *pratensis.* The flowers are white or pink. The sticky hairs sometimes function like flypaper and trap small insects, but the plant is not carnivorous.

7b Stem with glandular hairs only at the nodes; crown at center of flower small or lacking; flowers less than 1/4 inch (0.6 cm) broad (Fig. 176) .... SLEEPY CATCHFLY, *Silene antirrhina* L.

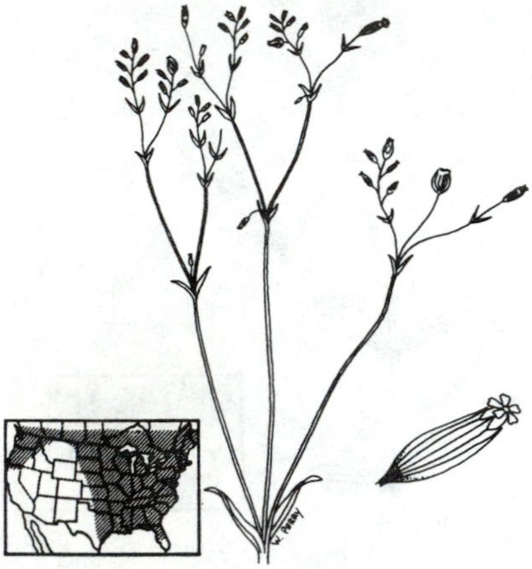

9a Leaves pointed at the tip, 1/4 inch (0.6 cm) long (Fig. 177) .... THYME-LEAVED SANDWORT, *Arenaria serpyllifolia* L.

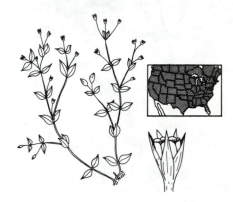

Figure 177

Figure 176

Sandy or gravelly soil. One of the native Silenes, these plants bloom in late spring and summer. The stems reach 35 inches (90 cm) with oblanceolate leaves at the base and lanceolate ones above. The corolla is white or pink but the notched petals are not much longer than the calyx. There is only a very small crown, or it is lacking altogether.

Sandy soil. The stems and leaves of this 1 foot (30 cm) plant are covered with tiny hairs. The leaves are ovate and less than 1/4 inch (5 mm) long. The petals are white and shorter than the sepals. As in most of the chickweeds this species has 10 stamens, but unlike *Cerastium*, there are usually only 3 styles.

8a (1) Petals 5, entire or with a shallow notch at the apex 9

8b Petals 5, deeply notched, or bifurcated so that there appear to be 10 petals 10

9b Leaves blunt at the tip, 1/2 to 1 inch (1.3-2.5 cm) long (Fig. 178) .... **BLUNT-LEAVED SANDWORT,** *Moehringia lateriflora* (L.) Fenzl

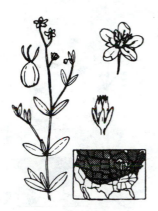

Figure 178

Synonym: *Arenaria lateriflora* L.

This species is found in moist places and gravelly shores in Alaska, the southern half of Canada, in the northern half of the United States, and in the Rocky Mountains to New Mexico. Also found in Europe and Asia. It is a delicate plant 2 to 8 inches (5-20 cm) high. The white flowers 1/4 to 1/3 inch (6-8 mm) broad have 3 styles and 8 to 10 stamens.

10a Styles usually 3 (genus STELLARIA) **11**

10b Styles usually 5 (genus CERASTIUM) **12**

11a Leaves ovate, the lower ones with petioles (Fig. 179) .... **COMMON CHICKWEED,** *Stellaria media* (L.) Vill.

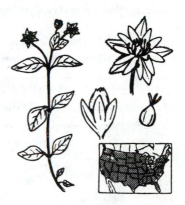

Figure 179

This plant, with spreading stems 2 to 16 inches (5-40 [-80] cm) long, is one of the most common weeds in North America and Europe and is found nearly everywhere—in cultivated areas, waste ground, and woodlands. It was introduced from Europe and is also native of Asia. The small star-like white flowers, 1/6 to 1/3 inch (4-9 mm) broad are produced from early spring to late fall, and even into winter in sheltered places with southern exposure. The stamens are 2 to 10 in number and each petal is deeply 2-parted. The ovate opposite leaves form a thick carpet when the plants are growing in dense patches, as is usually the case.

Since this herb is green early in the spring when other plants are still dormant it has had wide use fresh as a salad, or boiled or steamed as a green vegetable. The young stems and leaves are tender and with butter and salt are a tasty early spring wild harvest. The Indians also used the seeds (which are tiny) for bread or soup thickening. The seeds and shoots have been fed to chickens.

WATER MOUSE-EAR CHICKWEED,
Myosoton aquaticum (L.) Moench (*Stellaria aquatica* L.) is a similar appearing but somewhat larger European plant which grows in wet places throughout our area. It differs from *Stellaria media* in having 5 styles instead of 3.

11b Leaves narrow, less than 1/4 inch (6 mm) broad, linear to oblanceolate, sessile (Fig. 180) ....
.... **LONG-LEAVED CHICKWEED,**
Stellaria longifolia Muhl. ex Willd.

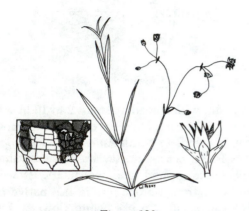

Figure 180

Moist meadows and woods. This perennial chickweed has 4-angled stems to 18 inches (45 cm) tall. There are many white flowers in an inflorescence, each flower with a long drooping stalk.

    Stellaria longipes Goldie is a similar spring-flowering plant growing only to 12 inches (30 cm) tall. It has fewer flowers, on erect pedicels. Both of these species grow worldwide in temperate and boreal regions.

12a Petals much longer than the sepals; stems not sticky (Fig. 181) ....
.... **FIELD CHICKWEED,**
Cerastium arvense L.

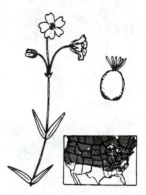

Figure 181

Dry rocky places throughout the northern hemisphere. The stems and leaves of this variable plant may be hairy or not. The stem reaches a height of 6 to 16 inches (15-40 cm) and bears pairs of linear leaves. The flowers are more than 1/2 inch (1.3 cm) broad, with white, two-lobed petals.

12b Petals shorter than the sepals or only slightly longer; stems sticky **13**

13a Pedicels shorter than calyx; petals shorter than the sepals or equalling them (Fig. 182) ....
....STICKY MOUSE-EAR CHICKWEED, *Cerastium glomeratum* Thuill.

14a Sepals with glandular hairs (Fig. 183) .... NODDING CHICKWEED, *Cerastium nutans* Raf.

Figure 183

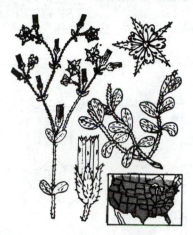

Figure 182

Synonym: *Cerastium viscosum* L.

This very common, viscid-pubescent European weed grows to a height of 4 to 12 inches (10-30 cm). The leaves are from 1/4 to 3/4 inch (0.6-2 cm) long and thickly coated with sticky hairs. The flowers are about 1/4 inch (6 mm) broad and often in tight clusters at the ends of the stem.

13b Pedicels longer than the calyx; petals equal to or longer than the sepals 14

The stems of this bright green annual are often reclining and may attain a length of 20 inches (50 cm). The whole plant is viscid-pubescent. The basal leaves are spatulate; those of the stem lanceolate. The flowers are about 1/4 inch (6 mm) across. As the fruits mature they begin to nod downwards from the pedicel.

SHORT-STALKED CHICKWEED, *C. nutans* var. *brachypodum* Engelm. & Gray, has much shorter pedicels. It ranges west from the Mississippi to the Rockies.

14b Sepals with long, non-glandular hairs (Fig. 184) .... COMMON MOUSE-EAR CHICKWEED, *Cerastium fontanum* Baumg. ssp. *triviale* (Link) Jalas

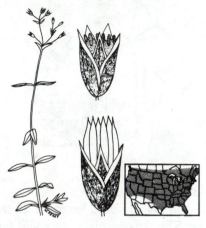

Figure 184

Synonym: *Cerastium vulgatum* L.

This Eurasian weed is spread over much of North America. The plant roots at the lower nodes of 6 to 20 inch (15-50 cm) stems and forms mats in lawns and open ground. The bracts of the inflorescence have membranous margins. The petals are about equal to, or longer than, the sepals.

NYMPHAEACEAE

WATERLILY FAMILY

1a Flowers yellow; sepals 5 or 6, yellow-green, forming a round cup; petals (a) small and stamen-like (Fig. 185) .... SPATTERDOCK, *Nuphar luteum* (L.) Sibth. & Sm.

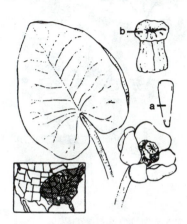

Figure 185

Synonym: *Nuphar advena* (Ait.) Ait. f.

In still or stagnant water in most of North America east of the Rocky Mountains. This attractive and common plant has showy yellow flowers, 1½ to 2 inches (3-5 cm) broad, with 6 sepals, many stamen-like petals (a), and numerous stamens. The compound pistil has 12 to 24 rays on the disk-like stigma (b), thus indicating the number of carpels.

The shiny green cordate leaves often float on the water. There are a number of varieties of this species based upon the size and shape of the leaves and the fruit shape. In some varieties the sepals have a red tinge or margin. ★ — FL

1b Flowers white or pink; sepals 4, green; petals 17 to 32, longer than the sepals (Fig. 186) ....
.... **FRAGRANT WATERLILY,** *Nymphaea odorata* Ait.

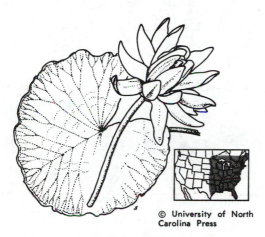

© University of North Carolina Press

Figure 186

These plants with nearly round shiny leaves notched at the base grow in ponds or slow moving streams. The flowers are large and striking, 2 3/4 to 4 3/4 inches (7-12 cm) broad, usually white, rarely pink, with many yellow stamens at the center. They open in the morning and become fragrant.

The LOTUS, *Nelumbo lutea* (Willd.) Pers., has a pale yellow flower with many sepals and petals and from a distance may be assumed to be a waterlily. However the flower itself is centered by a large inverse cone-shaped receptacle with ovaries sunken into it. The leaves are round, peltate and usually held above the water. The Sacred Lotus has pink or white flowers and has occasionally escaped from cultivation.

RANUNCULACEAE

BUTTERCUP FAMILY

In this family, identification of certain species depends upon knowing whether petals are present or not, a somewhat tricky circumstance because many of the sepals are brightly colored and petal-like. As a general rule in dicotyledons, if there is only one series of petal-like structures, these are in fact *sepals*. However, sepals sometimes fall away early, leaving only a series of *petals* in the mature flower. To be sure, examine the buds. Sepals are the outermost covering of the buds (except in Compositae). If the outermost covering is also the petal-like series, these are sepals.

There are more than 30 genera which comprise the large family Ranunculaceae, which numbers well over a thousand highly diversified and widely scattered species.

1a Flowers irregular (zygomorphic), with a single evident spur 2

1b Flowers regular (actinomorphic), if spurred, each petal or sepal (5) prolonged into a spur 4

2a Flowers white to pale blue; leaves mostly below the middle of the stem, divided into narrow segments (Fig. 187) .... PRAIRIE LARKSPUR, *Delphinium virescens* Nutt.

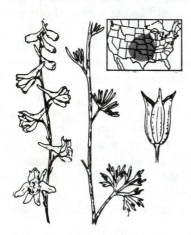

Figure 187

Prairies and open woods. The plant, 1 to 4 feet (30-122 cm) high, has compound leaves with narrow segments and a narrow raceme of pale blue or white flowers 1/2 to 3/4 inch (1.3-1.9 cm) long. The racemes are sometimes over 20 inches long. The 3 follicles are nearly erect. The spur in these flowers is formed by one of the 5 petal-like sepals. The 4 petals are smaller and the upper two form spurs that are enclosed within the sepal spur.

2b Flowers deep blue (white forms also exist); leaves either mostly basal and with oblong or wedge-shaped segments or leaves along the length of the stem and with linear lobes 3

3a Leaves few, mostly near the base, the leaf lobes broad; follicles widely spreading (Fig. 188) .... DWARF LARKSPUR, *Delphinium tricorne* Michx.

Figure 188

An early blooming plant, 1 to 2 feet (30-60 cm) high, of open woods. The flowers, 1 to 1½ inches (2.5-3.8 cm) long, are usually a deep bright blue in a loose raceme. Occasionally the flowers are white or light blue. The leaves are deeply 5 to 7 cleft. The 3 follicles are widely spreading in fruit.

Whoever devised the name "larkspur" knew both flowers and birds. Larks possess an unusually long spur or nail on the back toe. In these flowers the spur serves to keep the nectary so distant from the mouth of the flower that only long tongued insects can reach the nectar. This insures good pollination with a minimum expenditure of sweets.

3b Leaves along the length of the stem, the leaf divisions linear; follicles erect, each with a long beak (Fig. 189) ....
.... CAROLINA LARKSPUR, *Delphinium carolinianum* Walt.

Figure 189

Prairies and open woods. This somewhat pubescent slender stemmed plant attains a height of 32 inches (81 cm) and bears blue flowers. The tiny bracts are close to the calyx. The flowers are about an inch (2.5 cm) in length. The spur curves upward. The lower 2 petals are notched and hairy. The follicles are erect and each is tipped with a long beak.

All Larkspurs are poisonous. Grazing animals feeding on leaves and seeds have died. The plant parts contain alkaloids which act on the central nervous system of the animals (including humans) that eat them. Some people have reported rashes from touching the plants. Powdered seed is used in some remedies for lice.

4a Flowers with spurs 5

4b Flowers without spurs 6

5a Leaves compound; inflorescence a loose panicle; flowers nodding; petals spurred (Fig. 190) .... WILD COLUMBINE, *Aquilegia canadensis* L.

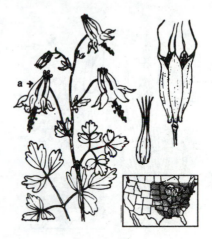

Figure 190

A common spring flower found in open rocky woods. The plant, 1 to 2½ feet (30-81 cm) high, has compound toothed leaves and scarlet and yellow nodding flowers 1 to 1½ inches (2.5-4 cm) long. Each of the 5 petals is spurred; the sepals are attached between the petals (a).

Although this is the only native Eastern Columbine, in the western states there are a number of species, some colored as the eastern species and others yellow, cream-colored, or combinations of blue and white.

5b Leaves entire, linear; flowers single on leafless scapes; sepals spurred (Fig. 191) .... MOUSETAIL, *Myosurus minimus* L.

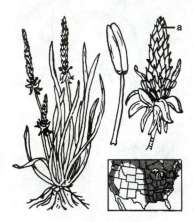

Figure 191

This tiny little annual grows in moist places and may reach a height of 1 to 6 inches (2.5-15 cm). The sepals have long spurs at the base and the greenish-yellow petals, when present, bear a claw. The stamens and pistils (a) are numerous. The pistils are borne on an elongate receptacle which makes the "tail."

6a Flowers clear yellow (white in related aquatic species); petals present or absent 7

6b Flowers greenish yellow, white, pink, lavender or blue; sepals usually petal-like 18

7a Petals absent; the sepals yellow and petal-like (Fig. 192) .... MARSH MARIGOLD, *Caltha palustris* L.

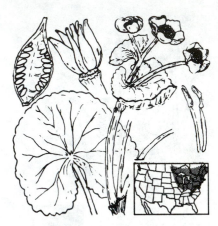

Figure 192

A common, stout plant, 8 to 24 inches (20-61 cm) high of swamps and wet meadows. The deep green, round or heart-shaped leaves are toothed and contrast strikingly with the brilliant yellow flowers which are over an inch (2.5 cm) broad. The stems are thick and hollow. These plants bloom early in the spring and are conspicuous when growing in large numbers.

Although the plant contains bitter and poisonous glucosides, it has been eaten—after boiling for an hour or more and changing the water at least once. For all the taste and food value left after that treatment it does not seem worthwhile to risk poisoning.

7b Petals present, large or very small, with a nectar pit at the base of the blade, yellow (also white in related aquatic forms not included in this key); sepals green (genus RANUNCULUS) 8

8a Basal and cauline leaves lobed or divided 9

8b Basal leaves, some or all, crenate 17

9a Plant aquatic; leaves divided into thread-like segments (Fig. 193) ....
.... YELLOW WATER-CROWFOOT,
Ranunculus flabellaris Raf.

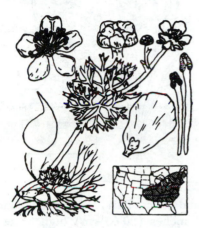

Figure 193

Submerged in shallow ponds or on muddy banks. The branching stems often become several feet long. The leaves which grow under water are very finely divided, while those developing above the water have expanded blades. The flowers are bright yellow, up to 1½ inches (3.8 cm) across, with 5 to 8 petals and numerous stamens and green pistils.

Another Water Crowfoot with yellow flowers is *R. gmelinii* DC. in which the leaf lobes are somewhat broader, to 3/16 inch (5 mm) wide and the petals less than 3/8 inch (8 mm).

There are also 4 species of aquatic Crowfoot with white flowers. These are mostly summer flowering but begin to bloom in May.

9b Plant of dry or marsh areas, but not normally submersed **10**

10a Petals shorter than or as long as the sepals, less than 1/4 inch (0.6 cm) long ....
.... **11**

10b Petals longer than the sepals, 1/4 to 5/8 inch (0.6-1.6 cm) long **12**

11a Stem glabrous; sepals spreading; achenes with hardly any beak (Fig. 194) ....
.... CURSED CROWFOOT,
Ranunculus sceleratus L.

from Stevens, 1910

Figure 194

Moist, low ground. The glabrous stems are up to 24 inches (60 cm) tall with doubly lobed lower leaves. The upper leaves are smaller and usually with three linear lobes. The flowers are small with short, pale yellow petals, 3/16 inch (5 mm) long or less and the many pistils arranged on a cylindrical receptacle in the center.

11b Stem hairy; sepals reflexed; achenes with a long hooked beak (Fig. 195) .... **HOOKED CROWFOOT,** *Ranunculus recurvatus* **Poir.**

from Stevens, 1910

Figure 195

Woods. This species is superficially similar to *Ranunculus scleratus* because the flowers have small pale yellow petals and the pistils constitute a green cylinder in the center of the flower. Here, however, the stem is slightly hairy and the pistils (achenes in fruit) have a long hooked beak.

Accurate identification of many of the crowfoots and buttercups depends on many characteristics of the achene (mature pistil), especially the overall shape and texture, and the shape and size of its margin and beak (persistent style).

12a Beak (the persistent style) of achene short, less than 1/2 as long as the body **13**

12b Beak of achene long and slender **15**

13a Base of plant swollen and bulblike (Fig. 196) .... **GOLDCUPS,** *Ranunculus bulbosus* **L.**

Figure 196

Fields and lawns. This hair-covered plant from 8 to 24 inches (20-60 cm) high arises from a bulblike base. The leaves are mostly basal and are basically 3-parted but each part is divided again so the leaves appear somewhat fernlike. The 5 to 7 petals are a brilliant yellow and make a flower that expands to a width of an inch (2.5 cm). This European native is also known as Meadow-gold.

13b Base of plant not bulbous **14**

14a Plant creeping, rooting at the nodes, somewhat hairy (Fig. 197) ....
.... CREEPING BUTTERCUP,
Ranunculus repens L.

14b Plant erect, hirsute (Fig. 198) ....
.... TALL BUTTERCUP,
Ranunculus acris L.

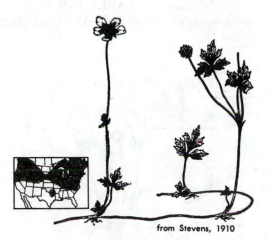

from Stevens, 1910

Figure 197

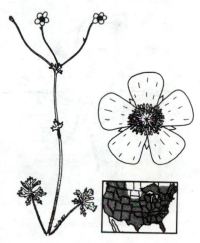

Figure 198

Fields, roadsides and lawns. The creeping habit of this plant can sometimes make it a pest. The leaves are 3-parted and the terminal lobe is stalked. All of the lobes may have smaller secondary lobes or teeth. The flowers are yellow with shiny petals up to 5/8 inch (1.6 cm) long and reflexed sepals. There is also a double flowered form.

Meadows and roadsides. A tall buttercup from Europe, this hairy plant can grow as high as 2 to 4 feet (60-122 cm). The leaves are mostly on the lower half of the stem. They are 3-lobed but the lobes are divided so that there appear to be five or more. The broadly ovate, enamel-yellow petals are twice as long as the sepals.

15a Roots slender; achene broadly margined(a); plant 1 to 2 1/2 feet (30-76 cm) high (Fig. 199)....SWAMP BUTTERCUP, *Ranunculus septentrionalis* **Poir.**

Figure 199

This buttercup grows 1 to 2½ feet (30-76 cm) high in low grounds and swampy areas. The dark green leaves are divided into 3 leaflets, each one usually stemmed. The flowers, 1 inch (2.5 cm) broad, are bright yellow with obovate petals. This buttercup blooms early and is quite common.

The petals of many of the buttercups have a characteristic high-glossed finish as though brushed with enamel paint.

15b Roots thickened; achene scarcely margined; plants 1 to 2 feet (30-61 cm) high **16**

16a Plant low, not over a foot (30 cm) high, pubescent with silky hairs; basal leaves appearing pinnate because the terminal segment is on a long stalk (Fig. 200) .... EARLY BUTTERCUP, *Ranunculus fascicularis* **Muhl. ex Bigelow**

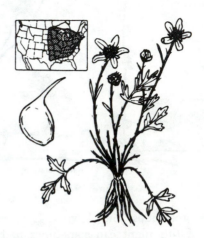

Figure 200

An early spring buttercup of hills and woodlands. The plant is from 6 to 12 inches (15-30 cm) high with deep yellow flowers about an inch (2.5 cm) broad. The dark green leaves are deeply 3 or 5 lobed and silky pubescent.

The family to which the buttercup belongs lends itself particularly well to making double flowers. Such flowers are usually the result of stamens becoming petal-like. Only a few families have sufficient numbers of stamens to make very double forms. ★ — TX

16b Plant up to 2 feet (60 cm) tall, densely villous when young; leaves mostly 3-lobed or 3-divided (Fig. 201) .... **HISPID BUTTERCUP,** *Ranunculus hispidus* **Michx.**

17a (8) Petals much longer than the sepals; basal leaves ovate; plant pubescent (Fig. 202) .... **DWARF BUTTERCUP,** *Ranunculus rhomboideus* **Goldie**

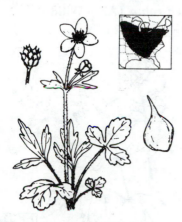

Figure 201

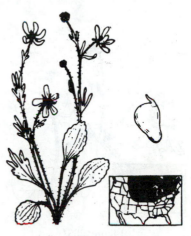

Figure 202

A buttercup of moist places and wooded areas. When young the plant, 8 to 24 inches (20-61 cm) high, is densely hairy, but is less so later in the season. The yellow flowers are 1/2 to 1½ inches (1.3-3.8 cm) broad.

 Most species of buttercup are considered poisonous to livestock, causing gastrointestinal upset and abortion. However, the plants are usually not eaten because of their bitter taste. In fact, a more tasty pasture plant is often assured of survival if it grows under the protecting leaves of a buttercup.

This is an attractive little prairie buttercup 4 to 8 inches (10-20 cm) high. The deep yellow flowers have fairly narrow petals less than 3/8 inch (5-9 mm) long. The basal leaves are ovate with crenate edges. The plant is covered with long silky hairs.

17b Petals small, about the same length as the sepals; basal leaves usually cordate; plant glabrous (Fig. 203) .... SMALL-FLOWERED CROWFOOT, *Ranunculus abortivus* L.

Figure 203

A common but rather inconspicuous plant of moist ground growing in most of the eastern half of North America. Very often it is found along a woodland stream. The plant is 6 to 20 inches (15-50 cm) high with pale yellow flowers about 1/3 inch (0.8 cm) broad, with small petals 1/8 inch (0.3 cm) long. The plant is glabrous and quite succulent. The leaves are bright green, the upper ones somewhat cut, the basal ones heart-shaped.

18a (6) Flowers many, in a terminal panicle or raceme; sepals (and petals, if present) deciduous so that the color of the flowers is mostly determined by the stamens 19

18b Flowers solitary, umbellate, or in a few-flowered panicle; petals absent; sepals petaloid (in *Hepatica* the involucre resembles a calyx) 22

19a Inflorescence a raceme; flowers white, perfect 20

19b Inflorescence a panicle; flowers greenish yellow, mostly unisexual 21

20a Raceme ovate; pedicels slender (Fig. 204) .... RED BANEBERRY, *Actaea rubra* (Ait.) Willd.

Figure 204

This is a common woodland plant 1 to 2 3/4 feet (30-82 cm) high. The tiny white flowers are in a dense terminal raceme. Because of the many stamens, the flower cluster has a feathery appearance. There are 4 to 10 white petals when the flower is young. The poisonous berry-like follicles, usually red, rarely white, are 3/8 inch (1 cm) long on slender pedicels. The leaves are compound.

Both this species and *Actaea pachypoda* have poisonous rhizomes, sap, and berries that can irritate the stomach and increase the pulse. The same species have been used against stomach troubles, dropsy and hysteria.

20b Raceme oblong; pedicels stout (Fig. 205) .... **WHITE BANEBERRY,** *Actaea pachypoda* Ell.

Figure 205

Synonym: *Actaea alba* (L.) Mill.

This baneberry grows in rich woods. It usually flowers a week or two later than *Actaea rubra*. The leaves are compound with toothed, ovate leaflets. The white flowers with narrow petals and deciduous sepals are on thick pedicels. The "berries" are usually china white with a black stigma scar. These berries give the plant the common name Doll's-eyes.

21a Leaves all with petioles; the leaflets with more than three lobes or teeth; stamens greenish yellow (Fig. 206) .... **EARLY MEADOW RUE,** *Thalictrum dioicum* L.

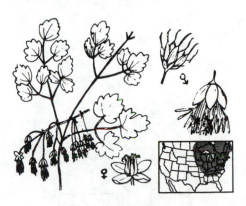

Figure 206

Common in thin woodlands. The staminate plant has an airy gracefulness about it because of the many long stamens. The plants are 1 to 2¼ (30-70 cm) feet high. The fruits in the genus *Thalictrum* are achenes.

This is a fairly large genus with almost 100 known species, many of which flower in the summer or fall. Several species are favorites for ornamental use and are often seen in borders.

21b Leaves on upper stalk sessile, the leaflets with 2-3 lobes or entire; stamens yellowish or purplish (Fig. 207) ....
.... WAXY MEADOW RUE, *Thalictrum revolutum DC.*

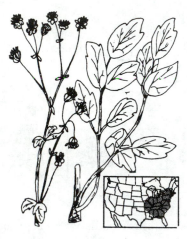

Figure 207

Woodlands and river banks. This plant may attain a height of 5 feet (1.5 m). The stems are heavy and often tinged with purple. The leaflets are barely lobed, thick, with obvious veins and a waxy coat beneath. The leaf margins are somewhat revolute. The plant flowers in late spring.

PURPLE MEADOW RUE, *Thalictrum dasycarpum* Fisch. & Lall., is a similar plant which also begins to bloom in late spring. It differs from the preceding in that the leaves are pubescent but not waxy-glandular.

22a Flowering stems with an involucral whorl of leaves below the flower stalks 23

22b Flowering stems with alternate leaves or small bracts but without an involucral whorl 29

23a Flowers in an umbel; involucre of 2 or 3 compound, sessile leaves with long-stalked oval or rounded leaflets; plant glabrous (Fig. 208) ....
.... RUE ANEMONE, *Thalictrum thalictroides* (L.) Eames & Boivin.

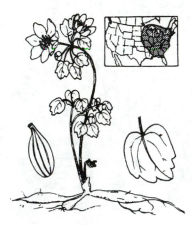

Figure 208

Synonym: *Anemonella thalictroides* (L.) Spach

A common, delicate spring flower of thin woodlands. The plant is 4 to 10 inches (10-25 cm) high with 3 white or pinkish-white flowers ½ to 1¾ inch (1.3-3.6 cm) broad. The thick tuberous roots and 4 to 15 achenes (1-seeded carpels) distinguish it from the False Rue Anemone, (Fig. 216), with which it is sometimes confused. The latter has (usually 4) 2 or 3 seeded pistils and fibrous roots (occasionally with tiny tubers).

23b Flowers not umbellate, either solitary on a single scape or 1 to 3 stalks arising at the involucre; involucre not of compound leaves, or if compound, the leaflets oblong with wedge-shaped bases; plants usually hairy 24

24a Leaves simple, strongly 3-lobed; involucre (a) of 3 leaves close under the flower, resembling a calyx (Fig. 209) .... HEPATICA, *Hepatica nobilis* P. Mill.

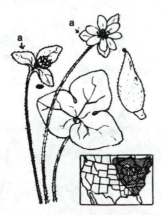

Figure 209

Synonym: *Hepatica americana* (DC.) Ker-Gawl.

An early-flowering woodland plant found in the southeastern part of Canada and northeastern United States. Also found in Europe, Alaska, and Asia. The plant is 4 to 6 inches (10-15 cm) high with blue, purple, lavender, or whitish flowers ½ to ¾ inch (1.2-2.5 cm) broad, borne singly on pubescent scapes. Another common name is Liverwort, so called because of the olive-green 3-lobed leaves. The leaves last through the winter and are the ones present when the plant blooms in the spring. The plant has been used medicinally in the past. There are two varieties of this species in North America. One, var. *obtusa* (Pursh) Steyermark, with obtuse leaf lobes, has previously been known as the species *Hepatica americana.*

24b Leaves compound or deeply cut or divided; involucre not calyx-like 25

25a Styles elongate, at maturity 3/4 to 1 1/2 inch (2-4 cm) long, densely plumose, flower to 2 inches (5 cm) broad (Fig. 210) .... PASQUE FLOWER, *Pulsatilla patens* (L.) P. Mill.

Figure 210

Synonym: *Anemone patens* L.

This charming, silky-hairy plant 4 to 16 inches (10-40 cm) high grows in dry prairie soil in Wisconsin and Illinois south to Texas and north-westward to British Columbia. It is also found in Europe and northern Asia. The blue or lavender (rarely white) flowers about 2 inches (5 cm) broad open early in March or April on rather short peduncles. The peduncle often grows as much as a foot (30 cm) by the time the achenes are mature. The basal leaves are many times divided into narrow lobes. The involucral leaves are also divided, but sessile. South Dakota has chosen this plant as the state flower.

25b Styles short, glabrous or pubescent; flowers less than 1 3/4 inches (4.5 cm) broad 26

26a Leaves of involucre sessile 27

26b Leaves of involucre petiolate 28

27a Flower solitary; achenes densely woolly (a); fruiting head of achenes cylindrical or ovate (b); plant of dry areas (Fig. 211) .... CAROLINA ANEMONE, *Anemone caroliniana* Walt.

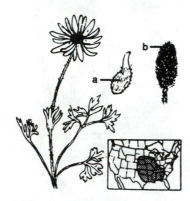

Figure 211

A colorful member of the Anemone genus found in open places. It grows 4 to 24 inches (10-60 cm) high with solitary purple, pinkish or nearly white flowers 3/4 to 1¾ inches (2-4.5 cm) broad. There are 5 to 20 petal-like sepals. The basal and involucral leaves are divided into 3 segments which are again divided, the involucral leaves sessile. The peduncle and sepals are villous. As the achenes mature, the fruiting head becomes cylindrical.

27b Flowering peduncles 1 to 3; achenes sparsely pubescent (a); fruiting head globose (b); plant of moist ground (Fig. 212) .... CANADA ANEMONE, *Anemone canadensis* L.

Figure 212

River banks and low ground. A coarse-stemmed, much-branched anemone 8 to 32 inches (20-81 cm) tall, with broad, sharply-toothed leaves. The white flowers are 1 to 1½ inches (2.5-4 cm) broad. The fruiting head is round.

This genus includes nearly 100 known species which are widely distributed throughout the temperate and arctic regions. Windflower is a name applied to many of them.

28a Plant delicate, 4 to 12 inches (10-30 cm) tall, nearly glabrous; flowers solitary; achenes with short hairs (Fig. 213) .... WOOD ANEMONE, *Anemone quinquefolia* **L.**

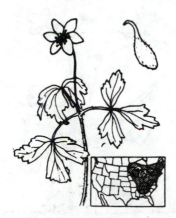

Figure 213

In thinly wooded areas. This common delicate plant is 4 to 12 inches (10-30 cm) high. The leaves are deeply lobed so that each leaf is composed of 3 to 5 segments, each with a wedge-shaped base. The solitary flowers are about 1½ inches (3.8 cm) broad, the 4 to 7 sepals usually white but often flushed with pink or rose on the outside. Each of the 15 to 20 carpels has a hooked beak. Another common name is Windflower.

28b Plant robust, to 3 1/4 feet (1 m) tall, pubescent; flowers 1 to 3 on stalks arising at the involucre; achenes with long hairs (Fig. 214) .... THIMBLEWEED, *Anemone virginiana* **L.**

from Stevens, 1910

Figure 214

Open woods. This tall anemone has pubescent stems and leaves. There are 1 to 7 basal leaves, each with 3 or 5 leaflets. The flowers are borne on long slender stalks, each of which may have a secondary involucre. The flowers are about 5/8 to 1 1/3 inches (1.5-3.5 cm) wide with (usually) 5 greenish-white sepals and densely hairy achenes.

29a (22) Sepals 3, deciduous; pistils numerous; fruit berrylike, red (Fig. 215) .... **GOLDEN-SEAL,** *Hydrastis canadensis* L.

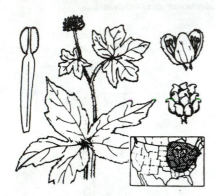

Figure 215

Rich woods. A plant 6 to 20 inches (15-51 cm) high from a thick yellow rootstock. There is usually one long-petioled basal leaf and two stem leaves. The solitary greenish-white flower is about ½ inch (1.3 cm) broad. There are 3 sepals but they fall off as soon as the flower opens. Since there are no petals, only stamens and pistils remain. The numerous stamens have broad filaments.

The yellow rhizome has had use for liver, stomach and urinary problems. It was used by the eastern Indians and listed in the U.S. Pharmacopoeia until the early 1900's.

29b **Sepals 5 or more, petal-like and persistent; pistils usually few; fruit a follicle** **30**

30a **Flowers with modified stamens (*staminodes*) present, solitary, yellowish or cream colored** **31**

30b Flowers without modified stamens present, 1 to few, pinkish to white (Fig. 216) .... **FALSE RUE ANEMONE,** *Isopyrum biternatum* Torr. & Gray

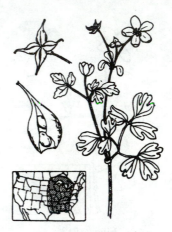

Figure 216

This delicate flower of moist woods and thickets resembles *Anemone quinquefolia* and *Thalictrum thalictroides*. Close inspection will show that the leaves are different and that the fruiting head consists of several dehiscent follicles instead of indehiscent achenes. The plant is 4 to 20 inches (10-51 cm) high with a few white flowers ½ to ¾ inch (1.3-2 cm) broad.

31a Plant 3 to 6 inches (7.5-15 cm) high; flowers whitish on naked or minutely bracted scapes; leaves 3-parted (Fig. 217) .... **GOLDTHREAD,** *Coptis trifolia* (L.) Salisb.

31b Plant 4 to 20 inches (10-50 cm) tall; flowers yellowish green; leaves palmately 5- to 7-parted (Fig. 218) .... **AMERICAN GLOBEFLOWER,** *Trollius laxus* L.

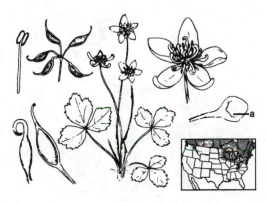

Figure 217

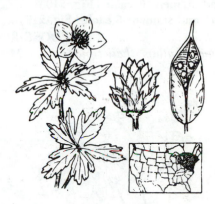

Figure 218

This is a dainty plant of mossy woods and swamps, and is found in Alaska, Greenland, Europe and Asia as well as in the region shown on the map.

The plant is 3 to 6 inches (7.5-15 cm) high with basal, 3-divided evergreen leaves and white flowers about ¾ inch (1.9 cm) broad, on naked scapes. The sepals are 5 to 7 in number; the staminodes (a) are 5 to 7, small and club shaped with a nectar pit at the tip. The name of Goldthread comes from the slender yellow rhizomes by which the plant spreads. The plant has been used as bitters in tonics.

This plant grows in swamps and wet ground. It is weak and often decumbent, with branches 5 to 20 inches (13-51 cm) long. The greenish-yellow flowers, about 1½ inches (3.8 cm) broad, are usually solitary. There are 5 to 7 petal-like sepals and 15 to 25 small staminodes. The fruiting head of follicles is nearly an inch (2.5 cm) broad. ★

BERBERIDACEAE

BARBERRY FAMILY

1a Flowers greenish, in a terminal panicle; leaves at least ternately compound; petals and stamens 6 each (Fig. 219) ....
(a) and stamens 6 each (Fig. 219) ....
.... **PAPOOSE-ROOT,**
Caulophyllum thalictroides (L.) Michx.

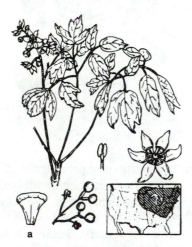

Figure 219

This early-flowering plant is found in rich woods. It is most common in its western range. The plant is 1 to 3 feet (30-91 cm) tall and has compound leaves and a panicle of yellowish-green flowers 3/8 inch (1 cm) broad. There are 6 petal-like sepals and 6 inconspicuous petals. After the ovary bursts, the two blue fleshy seeds remain exposed. Another common name is Blue Cohosh.

The roots and rhizomes had a brief official use in American medicine and a folk use as a diuretic and for menstrual problems.

1b Flowers white, solitary; leaves peltate or parted into 2 divisions; stamens 8 or more
.... **2**

2a Leaves with 2 equal divisions; flowers borne singly on naked scapes; petals and stamens 8 each (Fig. 220)....**TWINLEAF,**
Jeffersonia diphylla (L.) Pers.

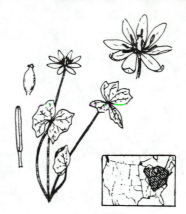

Figure 220

This woodland plant is 4 to 8 inches (10-20 cm) high when in flower but later reaches 16 or 18 inches (41-46 cm). The single white flower, about 1 inch (2.5 cm) broad, resembles *Sanguinaria canadensis* but is not as delicate or attractive. The leaves are deeply parted into 2 ovate divisions; this is the basis for the name Twinleaf. The leathery capsule is an inch (2.5 cm) long. The genus name was given in honor of Thomas Jefferson.

2b Leaves large, peltate, palmately lobed; flower single in the fork of a pair of leaves; petals 6 to 9; stamens usually twice the number of petals (Fig. 221) .... MAY APPLE, *Podophyllum peltatum* L.

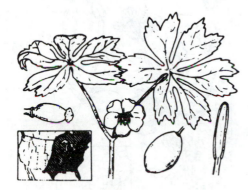

Figure 221

A common and conspicuous plant of low rich woods found in southeastern Canada and in much of the eastern half of the United States. The plant grows 1 to 1½ feet (30-46 cm) high with 2 large leaves and a single large waxy-white flower about 2 inches (5 cm) broad. The peltate leaves with 5 to 9 lobes are occasionally a foot (30 cm) in diameter. The flower is very attractive but has a heavy disagreeable odor. The edible fruit is a large greenish-yellow fleshy berry 1 to 2 inches (2.5-5 cm) long and is called Wild Lemon. Another familiar name is Wild Mandrake.

There are many references to the poisonous, edible, and medicinal properties of this plant. The active principal is podophyllum resin found in roots, foliage and the unripe fruits. Taken internally it causes diarrhea and vomiting. In small doses the ripe fruits are edible but somewhat laxative. The extremely caustic resin has had use, both folk and official, as a treatment for warts and tumors.

PAPAVERACEAE

POPPY FAMILY

1a Petals 8 to 12, white; leaves palmately lobed; juice red (Fig. 222) .... BLOODROOT, *Sanguinaria canadensis* L.

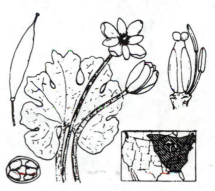

Figure 222

Rich open woods. The white flowers 1 to 2 inches (2.5-5 cm) broad on scapes up to 6 inches (15 cm) tall usually open before the leaves unfold and soon loose their 2 sepals and 8 to 16 petals. The leaves are attractive because of their heavy veins and uneven lobes. The plant has red juice, which makes a natural dye. The rhizomes and juice contain alkaloids which affect the stomach membranes and the nervous system. The juice is an old remedy for warts and skin cancers. Bees visit the flowers.

1b Petals 4, yellow; leaves pinnately lobed; juice yellow 2

2a Flowers terminal, about 1 inch (2.5 cm) broad; capsule oval, bristly (Fig. 223) .... CELANDINE POPPY, *Stylophorum diphyllum* (Michx.) Nutt.

2b Flowers in axillary umbels, 1/2 to 2/3 inch (1.3-1.7 cm) broad; capsule linear, glabrous (Fig. 224) ...... CELANDINE, *Chelidonium majus* L.

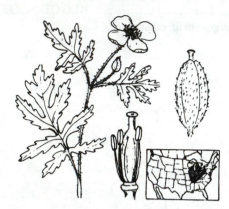

Figure 223

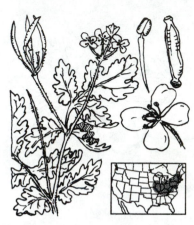

Figure 224

A woodland poppy which readily attracts attention, the plant, which is 12 to 20 inches (30-50 cm) high, has yellow juice and one to several large, golden yellow flowers. The petals are 3/4 to 1 1/8 inch (2-3 cm) long. The 2 sepals fall early. The bristly, ovoid capsule is about an inch (2.5 cm) long.

Moist soil. Native of Eurasia. This slightly pubescent, branching plant reaches a height of 1 to 3 feet (30-91 cm). The thin leaves, 4 to 8 inches (10-20 cm) in length, are frequently dilated at the base. The yellow flowers, a half inch (1.3 cm) or more across, are followed by slim, smooth fruiting capsules 1 or 2 inches (2.5-5 cm) long at maturity.

FUMARIACEAE

FUMITORY FAMILY

1a Each of the two outer petals spurred at the base (Fig. 226) (genus DICENTRA) 2

1b One of the two outer petals spurred at the base (Fig. 229) (genus CORYDALIS) 4

2a Inflorescence a panicle; flowers deep pink (Fig. 225) .... **WILD BLEEDING-HEART,** *Dicentra eximia* (Ker.) Torr.

3a Petal spurs triangular, diverging (Fig. 226) .... **DUTCHMAN'S-BREECHES,** *Dicentra cucullaria* (L.) Bernh.

Figure 225

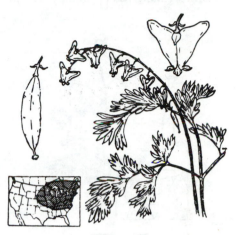

Figure 226

Found in rocky places in New York south in the mountains to Georgia and Tennessee. The plant grows 10 to 20 inches (25-51 cm) high with pink flowers about 1 inch (2.5 cm) long in a panicle. This is a wild cousin of the familiar Bleeding-heart, *D. spectabilis* Lem., of flower gardens.

There are several more species of Bleeding-hearts in the western mountains, all with fern-like dissected leaves and cordate flower bases. The colors range from yellow to cream to rose.

2b Inflorescence a raceme; flowers white, sometimes pinkish or greenish 3

Rich woods, especially on cliffs and river banks. The delicate, basal fern-like leaves are many times divided into narrow linear segments. The arching flower stalk is 4 to 12 inches (10-30 cm) tall with a terminal raceme of 3 to 12 nodding white or pinkish flowers held above the leaves. There are 2 small scale-like sepals and 2 pairs of petals. The outer two petals ½ to ¾ inch (1.3-2 cm) long are spurred at the base and enclose the 2 inner petals. There are 6 stamens, united into 2 sets of 3 each. The common name refers to the shape of the flowers, like so many pairs of tiny trousers hung out to dry. The seeds have an oily appendage that is attractive to ants, who consequently spread the seeds.

3b Petal spurs short and rounded, base of flowers cordate (Fig. 227) .... SQUIRREL CORN, *Dicentra canadensis* (Goldie) Walp.

4a Flowers pink with yellow tips (Fig. 228) .... PALE CORYDALIS, *Corydalis sempervirens* (L.) Pers.

Figure 227

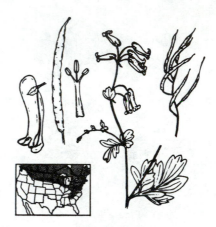

Figure 228

Rich woods. This plant is very much like *Dicentra cucullaria,* with the same dissected leaves and raceme of nodding flowers usually standing above the foliage. In this species the flowers are white or greenish white and the spurs are short and rounded. The small tubers at the base of the plant are the "corn" of the common name. They have been used in tonics but overdoses have caused convulsions in livestock.

Rocky places and open woods. The plant is 12 to 32 inches (30-80 cm) tall with pinnately dissected leaves. The flowers are in few-flowered racemes or panicles. The yellow-tipped pink flowers, 3/8 to 3/4 inch (10-17 mm) long, like all Corydalis, have 2 sepals and a spurred upper petal. In this species the spur is short and rounded. The narrow capsule is 3/4 to 1½ inches (2-4 cm) and tipped with the persistent style.

The species also occurs in Alaska.

4b Flowers totally yellow **5**

5a Outer petals keeled (a), not crested (Fig. 229) .... **GOLDEN CORYDALIS,** *Corydalis aurea* Willd.

6a Flowers pale yellow, leaves usually 3/4 to 1 1/4 inch (2-3 cm) long (Fig. 230) .... **SLENDER FUMEWORT,** *Corydalis micrantha* (Engelm.) Gray

Figure 229

Rocky soil on banks and in open woods. This species is 8 to 20 inches (20-50 cm) high with finely dissected leaves and a short, dense raceme of bright yellow flowers about ½ inch (1.3-1.6 cm) long. Each of the outer petals is folded into a keel. The spur is about ¼ inch (4-5 mm) long. The knobby elongate capsules hold a row of shiny black seeds. Another common name is Golden Fumeroot.

5b Outer petals with a conspicuous crest (Fig. 231-a) **6**

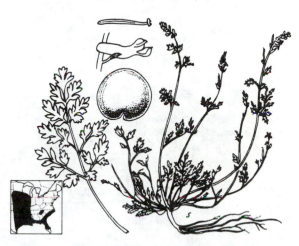

© University of North Carolina

Figure 230

Moist, sandy soils. This glaucous or green plant grows up to 16 inches (40 cm) tall. It has small, doubly compound leaves. The racemes of flowers are much taller than the leaves. The spurred petal is 3/8 to 5/8 inch (0.9-1.5 cm) long with the spur about half of the length of the petal. The elongate capsule is smooth.

6b Flowers bright yellow; leaves 1 1/8-3 1/8 inch (2.9-8 cm) long (Fig. 231) .... VESICULAR CORYDALIS, *Corydalis crystallina* Engelm.

Figure 231

A stout attractive *Corydalis* growing on prairies and fields of the Midwest states. The plant is 8 to 20 inches (20-51 cm) high with finely-cut leaves and bright yellow flowers, 1/2 to 2/3 inch (1.3-1.7 cm) long, in several to many racemes. Only one of the outer petals is spurred. The spur is 1/4 to 1/3 inch (0.6-0.8 cm) long. Both outer petals are conspicuously crested (a). The capsule is 3/4 inch (1.9 cm) long and densely covered with transparent vesicles which make the fruit appear rough.

CRUCIFERAE [BRASSICACEAE]

CABBAGE FAMILY [MUSTARD FAMILY]

The Cabbage Family is a well-differentiated, homogeneous family. All members have flowers with parts in the cross-shaped arrangement that gives the family its name. The stamens are characteristically of 2 lengths, 4 long and 2 short (or occasionally reduced to 2 or 4 stamens of equal length). The leaves are often variably toothed and divided even on the same plant. Therefore, since the flowers are so similar and the leaves so variable, identification relies heavily on fruit characters as well. Fortunately fruits of the Cruciferae develop rapidly and often immature fruits are present on the inflorescence even while the upper flowers are still in bud. If fruits are not present it is still often possible to look at the ovary and discover whether it is long or short, and thus extrapolate somewhat the shape of the fruit.

1a Petals with yellow blade (broadened upper portion) 2

1b Petals with white or purplish blade (sometimes yellowish at the base) 8

2a Fruits nearly globose (Fig. 232) .... FALSE FLAX, *Camelina microcarpa* Andrz.

Figure 232

Naturalized from Europe in fields and disturbed sandy soil. This plant with pale yellow to creamy flowers has pubescent stems from 1 to 2 feet or more (30-70 cm) tall. The rosette leaves, when present at flowering, are oblanceolate, toothed and pubescent. The leaves on the stem may be pubescent or glabrous and are lanceolate to linear and sessile with the bases pointed backward into ear-like projections which clasp the stem. The petals are about 1/8 inch (3 mm) long surrounding an oval or pear-shaped ovary. The mature fruit is obovoid or pear shaped and, including the beak (persistent style), is slightly more than 1/4 inch (6 mm) long.

Camelina sativa (L.) Crantz is also established but less commonly scattered over the United States and Canada. It is nearly glabrous, with fruits to 3/8 inch (1 cm) long.

2b **Fruits narrow and elongate** 3

3a **Leaves cordate-clasping, entire; petals pale yellow (Fig. 233) .... HARE'S-EAR,** *Conringia orientalis* **(L.) Dumort.**

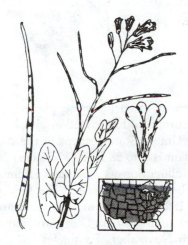

Figure 233

This European immigrant has become a bad weed in some areas. It is light green and grows to a height of 1 to 2½ feet (30-80 cm). The leaves, from which it gets its common name are cordate, slender and up to 4 inches (10 cm) long and 1¾ inches (4.5 cm) wide. The petals are pale yellow to creamy white with

a length up to 3/8 inch (1 cm) long. The fruit becomes 3 to 5 inches (7.5-13 cm) long at maturity and is 4-angled. Another name is Treacle Mustard.

3b **Leaves not cordate, usually toothed or dissected** **4**

4a **Flowers about 3/4 inch (1.9 cm) across; lower leaves entire or dentate, not deeply lobed (Fig. 234)** .... **WESTERN WALLFLOWER,** *Erysimum asperum* **DC.**

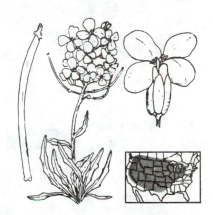

Figure 234

An attractive plant 8 inches to 3¼ feet (20-100 cm) high found in open places and thin woods. The fragrant yellow flowers about 3/4 inch (1.9 cm) wide with petals 5/8 to 1 inch (1.5-2.5 cm) long are in a raceme which is sometimes 3 inches (7.5 cm) across. The stem leaves are linear or oblanceolate and are usually dentate. The plant is rough-pubescent. The pods are sometimes 4 inches (10 cm) long. ★ —AK, Yukon

WORMSEED MUSTARD, *Erysimum cheiranthoides* L., blooms in late spring and through summer as far south as North Carolina in the East, and Colorado to Oregon in the West. It has smaller flowers with petals less than 1/4 inch (5.5 mm).

4b Flowers from 1/16 to 1/2 inch (1.6-13 mm) broad; lower leaves deeply lobed or dissected 5

5a Lower leaves with 1 to 4 pairs of ovate divisions, the terminal lobe round to broadly ovate with irregularly sinuate margins; upper stem leaves auriculate clasping (Fig. 235) .... YELLOW-ROCKET, *Barbarea vulgaris* R.Br.

5b Lower leaves with lanceolate divisions, the terminal lobe or blade coarsely toothed; upper stem leaves not auriculate clasping 6

6a Upper stem leaves with a pair of linear divisions and a hastate terminal lobe; silique gradually acuminate (Fig. 236) .... **HEDGE MUSTARD**, *Sisymbrium officinale* (L.) Scop.

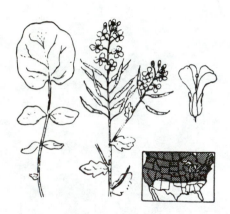

Figure 235

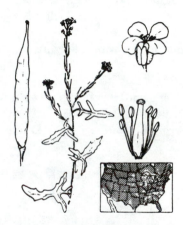

Figure 236

Another plant that was introduced from Europe and is now established in fields and along roadsides in most of North America. The sturdy erect plant is 1 to 2¾ feet (30-80 cm) high with bright yellow flowers 1/3 inch (8 mm) broad and pinnatifid lower leaves. The stem leaves gradually become smaller upwards on the stem and the terminal lobe less rounded and more toothed. The uppermost leaves are without small divisions, sessile, auriculate, and clasping. The fruit (silique) is linear, 5/8 to 1 1/8 inch (1.5-3 cm) long. Another name is Yellow Cress. It is an early spring flowering plant and is sometimes used for greens. As with many members of the family it is an excellent source of Vitamin C.

A common weed of waste places introduced from Europe and now well-established in most of the United States except in the Northwest. The plant is 1 to 2½ feet (30-76 cm) high with small yellow flowers 1/16 inch (2 mm) broad, and lobed, toothed leaves. The seed pods are about 1/2 inch (1.3 cm) long and are closely pressed to the stem. The plant blooms from May to November. Young leaves and shoots are gathered for cooked greens which are well stocked with vitamin C. Older plants eaten in quantity have caused illness in livestock.

6b Upper stem leaves entire, with toothed or entire margins 7

7a Flowers 3/8 inch (1 cm) wide; beak of pod short (Fig. 237) .... **BLACK MUSTARD,** *Brassica nigra* (L.) W.D.J. Koch

7b Flowers slightly more than 1/2 inch (1.5 cm) wide; beak of pods at least 1/3 as long as the body; pod bristly (Fig. 238) .... **WHITE MUSTARD,** *Sinapis alba* L.

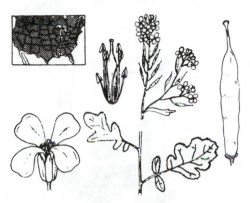

Figure 237

Figure 238

This plant, the source of commercial mustard seed, was introduced from Europe and is now established over most of the United States. It is native in Asia. The branched plant is 1 to 5 feet (30-150 cm) high with yellow flowers about 3/8 inch (1 cm) broad. The pods are erect and usually pressed close to the stem. They are 3/4 inch (2 cm) long with a short beak. Mustard seed is used as a flavoring in foods and in quantity to induce vomiting. On the skin it produces heat and reddening. The greens, although the lower leaves are hairy, can also be used in salads and as a vegetable.

LEAF MUSTARD, *Brassica juncea* (L.) Czern., has also escaped and blooms in the South in spring. It has similar uses to Black Mustard. The leaves are not hairy and are somewhat glaucous. The pods are slightly spreading from the stem.

Brassica is an important genus which contains several valuable food plants. Cabbage and Brussel Sprouts (*Brassica oleracea* L.), Turnip (*B. rapa* L.) and Rutabaga (*B. napus* L.) are garden plants which escape and begin to bloom in spring.

Synonym: *Brassica hirta* Moench

This mustard grows from 1 to 2 feet (30-60 cm) with lower leaves up to 8 inches (20 cm) long. The stem leaves are irregularly pinnately dissected. The yellow flowers are about 1/2 inch (1.5 cm) wide. The ovaries and bodies of the fruits are bristly. The beak is glabrous and may be 1 or 2 times as long as the body.

CHARLOCK, *Sinapis arvensis* L., is a similar weed except that the ovaries and fruits are not bristly and the beak may be only 1/3 as long as the body of the mature fruit. Synonym: *Brassica kaber* (DC.) Wheeler.

8a **(1)** Plants 8 inches (20 cm) tall or less; leaves simple, with stiff stellate hairs, basal or mostly near the base 9

8b Plants usually taller than 8 inches (20 cm); leaves compound or deeply lobed, or simple and borne the full length of the stem 10

9a Petals 2-cleft; pods short, oval; leaves basal only (Fig. 239) .... **VERNAL WHITLOW-GRASS**, *Erophila verna* L.

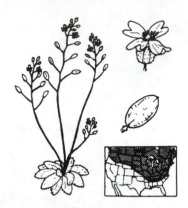

Figure 239

Synonym: *Draba verna* L.

A tiny plant 2 to 8 inches (5-20 cm) high of sandy areas and roadsides introduced from Europe and now established in much of the United States. The oblanceolate, stellate-pubescent leaves are in a basal rosette commonly 3/8 to 3/4 inch (1-2 cm) broad. From the rosette arise 1 to several leafless slender scapes with racemes of tiny white flowers 1/8 inch (3 mm) broad. The flowers have 2-parted petals and ovoid ovaries.

9b Petals entire (occasionally absent); pods considerably longer than wide; leaves basal and on a short stem (Fig. 240) .... **CAROLINA WHITLOW-GRASS**, *Draba reptans* (Lam.) Fern.

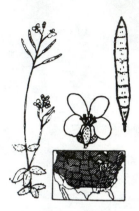

Figure 240

This, being a winter annual, flowers very early. The total height of the plant is 2 to 8 inches (5-20 cm) but leaves are only borne on the lower quarter. They are elliptic to obovate, entire, 1 1/8 inches (3 cm) long and densely pubescent with stellate hairs beneath and mostly simple hairs above. The white flowers are in short racemes at the ends of slender leafless scapes. The white petals are entire or with a slight notch, less than 1/4 inch (0.6 cm) long. The oblong pods may become 1/2 inch (1.3 cm) long. It is usually found in sandy fields and woods.

WEDGE-LEAVED WHITLOW-GRASS, *Draba cuneifolia* Nutt. ex Torr. & Gray, is another winter annual which flowers very early. It ranges from Illinois to California and Mexico and may be recognized by its wedge shaped, somewhat dentate leaves and pubescent fruits and pedicels. SHORT-FRUITED WHITLOW-GRASS, *Draba brachycarpa* Nutt. ex Torr. & Gray, with a range through the Southeast and into Kansas and Texas, has stems that are leafy nearly to the base of the inflorescence, and fruits about 1/8 inch (3-5 mm) long.

10a Leaves palmately divided into 3 lobes, at or above the middle of the stem and from the rhizome (genus DENTARIA) **11**

10b Leaves entire or pinnately lobed **12**

11a Flowers white; stem leaves 2, each with 3 ovate divisions (Fig. 241) .... TWO-LEAVED TOOTHWORT, *Dentaria diphylla* Michx.

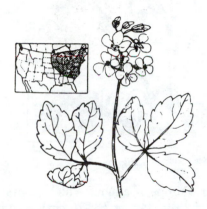

Figure 241

Rich woods and damp meadows. The smooth plant, 8 to 16 inches (20-40 cm) high, has long-petioled basal leaves and generally 2 stem leaves. The white flowers with petals 3/8 to 5/8 inch (1.0-1.7 cm) long are in a short terminal raceme. The pods are an inch (2.5 cm) or more in length. Another name is Crinkle-root, because the wrinkled edible rhizome tastes like pungent watercress.

LARGE TOOTHWORT, *D. maxima* Nutt., often has 3 stem leaves, 2 of them somewhat paired. The lateral lobes of the leaves are usually 2-lobed. It ranges from Maine to Michigan and Kentucky but is not common.

11b Flowers lavender or whitish; stem leaves usually 3, each with 3 narrow divisions (Fig. 242) .... CUT-LEAVED TOOTHWORT, *Dentaria laciniata* Muhl.

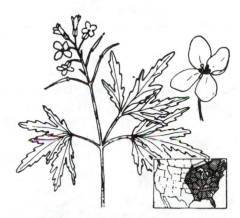

Figure 242

Rich damp woods, especially in flood plains. This Toothwort grows 8 to 16 inches (20-40 cm) high and has a raceme of white or lavender-pink flowers 1/2 to 3/4 inch (1.3-1.9 cm) broad. Usually there are 3 leaves semi-whorled above the middle of the stem. The leaves are 3-parted and each part is lobed or toothed; the lateral divisions may be deeply 2-lobed so that the leaf appears 5 parted. There are similar leaves from the rhizome but they are not present at flowering time. The pods are 1 to 1½ inches (2.5-4 cm) long. The whole plant arises from a tuberous, fusiform rhizome. Bees visit these flowers.

SLENDER TOOTHWORT, *D. hetero-phylla* Nutt., has broad basal leaves as in *D. diphylla* but stem leaves narrowly divided as in *D. laciniata*. It ranges from New Jersey and Pennsylvania south through the mountains to Georgia. The flowers are purplish.

12a Pods short and flat (much thinner than wide) less than 4 times as long as wide, divided by a septum (visible as a vein or channel) at right angles to the flat side of the fruit **13**

12b Pods 4 to many times as long as wide, if flattened, the septum parallel to the flattened sides **16**

13a Pods wedge-shaped (a); plants 6 to 20 inches (15-51 cm) high (Fig. 243) .... SHEPHERD'S PURSE, *Capsella bursa-pastoris* (L.) Medic.

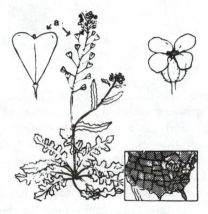

Figure 243

A very common weed that has been naturalized from Europe and is now distributed in fields and waste areas over the entire world. The plant, 6 to 24 inches (15-61 cm) high, has small white flowers about 1/8 inch (4 mm) long in one to many long racemes. The basal leaves are variously lobed and pinnatifid and usually form a rosette close to the ground. The pods are triangular and are noticeable later in the season when there are many of them. This plant can be found blooming almost any month of the year. As a green, leaves make a mild vegetable or salad. Medicinally the herb was used as an astringent and a diuretic.

13b Pods rounded (disk shaped) **14**

14a Pods broadly winged at the top and sides, with several seeds in each of the 2 cells, pods nearly 1/2 inch (1.3 cm) across when mature (Fig. 244) .... FIELD PENNY-CRESS, *Thlaspi arvense* L.

Figure 244

Another European annual which grows to a height of 1¼ to 2½ feet (38-76 cm). The basal leaves are petiolate but die early; the upper stem leaves are sessile with an acutely auricled base. The small pure-white flowers stand at pleasing contrast with the brilliant green foliage of young plants. Fall seedlings often bloom very early in the spring.

Although fruits and seeds reportedly cause illness in cattle, the plants have been cultivated in Europe and collected from the wild for young shoots to use as salad or cooked greens. It is bitter, but high in vitamins and sulphur.

PERFOLIATE PENNY-CRESS, *Thlaspi perfoliatum* L., is a smaller plant with leaves clasping by rounded auricles.

14b Pods narrowly winged across the top, with one seed in each of the two cells of the pod (genus LEPIDIUM) 15

15a Stem leaves tapered to the base; stamens 2 or 4 (Fig. 245) ....
.... **WILD PEPPERGRASS,**
Lepidium virginicum L.

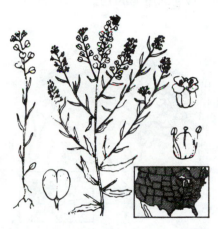

Figure 245

Common along roadsides and waste places in most of North America. The plant, 4 to 20 inches (10-51 cm) high, bears small white flowers about 1/12 inch (2 mm) broad in one to many racemes up to 4 inches (10 cm) long in flower and 10 inches (25 cm) in fruit. The basal leaves are obovate or spatulate and are usually lobed. The stem leaves are linear or lanceolate, entire or sharply toothed. Most of the Cruciferae have 6 stamens, but this species has only 2 or 4. Occasionally the petals are absent. The seed pod is nearly round and slightly winged above, about 1/8 inch (3-4 mm) long. Seeds gathered by the Indians in California were parched, ground, and used as mush. As indicated by an alternate common name "Poor-man's-pepper," the seeds are used as seasonings.

15b Stem leaves auriculate; stamens 6 (Fig. 246) .... **FIELD CRESS,**
Lepidium campestre (L.) R. Br.

Figure 246

This pubescent plant of fields, roadsides, and barnyards reaches 8 to 20 inches (20-50 cm) high. The basal leaves are 1½ to 3 inches (4-8 cm) long, oblanceolate, and toothed or pinnately lobed. The stem leaves are lanceolate and entire or toothed and auriculate. The flowering racemes are dense with white flowers, and about 6 inches (15 cm) long.

The mature pods are oblong to ovate, with a conspicuous wing at the top, somewhat curved (scoop shaped) and about 1/4 inch (6 mm) long. Another name is Cow Cress.

16a Plant aquatic, floating in clear streams or springs, or creeping on the banks, rooting at the nodes (Fig. 247) .... TRUE WATERCRESS, *Nasturtium officinale* R. Br.

Figure 247

Found in brooks, ditches, and streams in southeastern Canada, and in much of the United States. It was introduced into the United States from Europe and is now naturalized. The tiny white flowers are 1/4 inch (6 mm) broad, later developing pods 1/2 to 1 inch (1.3-2.5 cm) long. The plant bears roots at the nodes. The leaves have a pungent, refreshing taste and are used in salads and for garnishing. They are succulent and shiny and deeply pinnately divided.

PENNSYLVANIA BITTER-CRESS, also known as Native Watercress, *Cardamine pensylvanica* Muhl. ex Willd., grows with the true Watercress and in swamps and wet woods. It is similar to *Nasturtium officinale* but tends to be more erect and the lateral leaf divisions, especially on the upper leaves, are commonly linear and the terminal lobe lanceolate or wedge shaped. There is a single row of smooth seeds in the pod rather than a double row of rough seeds as in *Nasturtium*.

16b **Plant terrestrial** **17**

17a Middle and upper stem leaves entire .. 18

17b Middle and upper stem leaves pinnately lobed 26

18a Flowers large, petals 3/4 to 1 inch (2-2.5 cm) long (Fig. 248) .... DAME'S-ROCKET, *Hesperis matronalis* L.

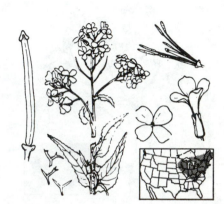

Figure 248

This 2 to 5 feet (0.6-1.5 m) tall species is an attractive member of the Mustard Family that is a native of Europe and Asia and has escaped from cultivation in the United States. The fragrant flowers, 2/3 to 1 inch (1.7-2.5 cm) broad, are purplish or white in a large raceme. The leaves sometimes reach 8 inches (20 cm) in length. The upper ones are ovate or ovate-lanceolate and sessile, and the lower leaves lanceolate to oblanceolate on petioles. The leaves are toothed and pubescent with forked hairs. Another name is Dame's Violet.

18b **Flowers small to medium, petals 1/8 to 5/8 inches (0.3-1.6 cm) long** **19**

19a Leaves broadly cordate to triangular; foliage with garlic-onion odor (Fig. 249) .... GARLIC MUSTARD, *Alliaria petiolata* (Bieb.) Cavara & Grande

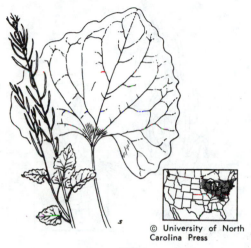

Figure 249

Synonym: *Alliaria officinalis* Andrz.

Moist woods, especially flood plains, and roadsides. This European native grows to 3¼ feet or more (1 m) tall with glabrous upper stems. The leaves are all petiolate with toothed, usually crenate, margins. The lower leaves are kidney shaped and gradually become triangular upwards. The white flowers with elongate ovaries and petals 1/4 inch (5-6 mm) long are in short broad, terminal racemes. Racemes in fruit are much longer. The pods are 4-angled, erect or slightly spreading, up to 2 3/8 inches (6 cm) long at maturity.

The leaves add flavor to salads, sandwiches and stuffings for pork. An old English name is Sauce-alone. The herb and seeds were also used in medicines.

19b Leaves lanceolate oblong or linear on the middle and upper stem; foliage without onion odor **20**

20a Lower leaves pinnately-lobed with 1-4 pairs of small lobes below a large terminal lobe, longer than 1 1/2 inches (4 cm); petals violet to white, the blades obovate-triangular; filaments flattened; fruits round in cross section (Fig. 250) .... PURPLE-ROCKET, *Iodanthus pinnatifidus* (Michx.) Steud.

Figure 250

This glabrous plant, growing to a height of 1 to 3 feet (30-91 cm) has many violet colored flowers (sometimes white) borne in an open panicle. The leaves are dentate and sometimes have several pinnate lobes near the petiole. The petioles often clasp the stem. The narrow pods may be an inch (2.5 cm) or more in length and usually stand erect.

20b Lower leaves entire, or if pinnately lobed and present at flowering, the basal leaves less than 1 1/2 inches (4 cm); filaments slender; fruits at least somewhat flattened **21**

21a Lower leaves lanceolate, entire or pinnately lobed 22

21b Lower leaves ovate or orbicular; base of stem tuberous 25

22a Basal leaves pinnately lobed; stems several from the rosette, spreading (Fig. 251) .... LYRE-LEAVED ROCK-CRESS, *Arabis lyrata* L.

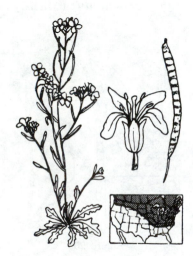

Figure 251

A plant, 4 to 16 inches (10-40 cm) high, of rocky and sandy places. Several branches arise from the basal rosette. The white flowers are up to 3/8 inch (1 cm) long, the petals spreading. The rosette leaves are 1½ inch (4 cm) long and pinnately divided with 2 to 4 pairs of lateral lobes. The lower stem leaves are toothed, the upper ones entire. This species also grows in Japan and Alaska.

22b Basal leaves entire; stems usually single, erect 23

23a Upper stem leaves tapered to a petiole; fruits 1/8 inch (3 cm) broad, strongly curved, on reflexed pedicels (Fig. 252) .... SICKLEPOD, *Arabis canadensis* L.

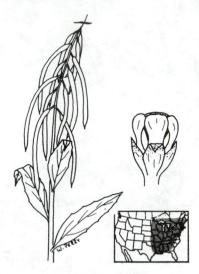

Figure 252

Woods. The stems are up to 3¼ feet (1 m) tall, bearing dentate leaves which narrow to the base and are without auricles. The flowers are white to pink with petals less than 1/4 inch (6 mm) long. The plant is most conspicuous in fruit when the curved, flattened pods are on reflexed pedicels.

23b Upper stem leaves auriculate-clasping 24

24a Pods spreading, curved; plant glabrous (Fig 253) ...... SMOOTH ROCK-CRESS, *Arabis laevigata* (Muhl.) Poir.

24b Pods erect; plant pubescent (Fig. 254) .... HAIRY ROCK-CRESS, *Arabis hirsuta* (L.) Scop.

Figure 253

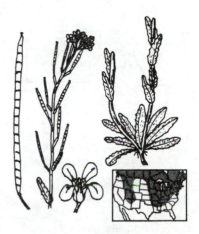

Figure 254

This plant may reach a height o 3¼ feet (1 m). The whitish bloom covering the leaves and stems lends a characteristic appearance. The basal leaves are spatulate and petiolate, but the stem leaves are sessile and have clasping auricles. The lower ones are toothed but the upper ones are entire. It grows in rocky woods. The flowers are greenish-white. The curved, spreading fruits are less than 1/8 inch (3 mm) wide.

Rocky places. This Rock-cress is usually roughly hairy although it may occasionally be nearly glabrous. The stems grow to 32 inches (80 cm) high. The flowers are very small and white, greenish white, or pink. The pods are 1 to 2 inches (2.5-5 cm) long, very slender and flattened, and usually held erect. They bear one row of seeds in each compartment.

TOOTHED ROCK-CRESS, *Arabis shortii* (Fern.) Gleason, is also roughly pubescent; it differs from the above in having spreading pods and more heavily toothed leaves. Its range is much the same.

TOWER MUSTARD, *Arabis glabra* (L.) Bernh., also has erect fruits. The stem is pubescent toward the base, and glabrous toward the top. The auriculate leaves are glabrous and glaucous. Its fruits are round in cross section, a character which helps distinguish this species from *A. hirsuta*.

25a Petals purple or lavender; sepals purple; .... PURPLE CRESS, *Cardamine douglassii* Britt.

25b Petals white; sepals green; stem leaves 5 to 10 (Fig. 256) .... SPRING CRESS, *Cardamine bulbosa* (Muhl.) B.S.P.

Figure 255

Figure 256

Found in rich damp woods in most of Canada east of the Rockies and in the United States from Maryland and Kentucky to Wisconsin. It is more common in the northern part of its range. The slender plant, 8 to 16 inches (20-40 cm), high has a raceme of showy purplish, or rarely white, flowers 1/2 to 2/3 inch (1.3-1.7 cm) broad and long-petioled ovate or orbicular basal leaves. The stem leaves are mostly sessile and are entire or have dentate margins. The nearly erect pods are about an inch (2.5 cm) long.

A delicate member of the Mustard Family found in wet meadows of southeastern Canada and in the eastern half of the United States. The glabrous plant grows 8 to 24 inches (20-61 cm) high with white flowers, rarely pinkish, about 1/2 inch (1.3 cm) broad in a raceme. This species flowers several weeks later than *Cardamine douglassii* and is found much farther south. The leaves and seed pods of both species are quite similar.

The plant, especially the tuberous rootstock, is a substitute for horseradish.

26a (17) Flowers 1/4 to 1/2 inch (6-13 mm) broad (Fig. 257) .... **MEADOW BITTER-CRESS,** *Cardamine pratensis* L.

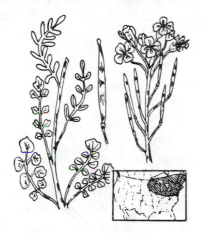

Figure 257

This species is native in cool regions around the northern hemisphere. The stems are 8 to 20 inches (20-50 cm) tall. The basal leaves have broad and dentate divisions as do the lowest stem leaves. The upper leaves have oval to oblong leaflets.

26b Flowers 1/8 inch (3-4 mm) broad (Fig. 258) .... **SMALL-FLOWERED BITTER-CRESS,** *Cardamine parviflora* L.

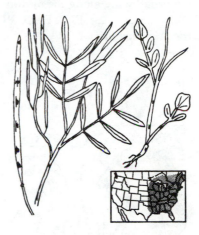

Figure 258

This is a slender plant of fields and woods, blooming from March to May. The stems are up to 10 inches (25 cm) tall. The leaves on the stem are 3/4 to 1½ inches (2-4 cm) long and have very narrow segments, less than 1/8 inch (2 mm) wide. The basal leaves have round or obovate segments. The white, spatulate petals are at most about 1/8 inch (3 mm) long.

HAIRY BITTER-CRESS, *C. hirsuta* L., has many basal leaves. The petioles have long hairs at the base, and the blades are finely hairy on the top. The flowers are very small. The plant is common in the Old World and naturalized in the eastern United States.

SARRACENIACEAE

PITCHER PLANT FAMILY

1a Flower reddish brown; leaves urn shaped, with a broad wing along the midline (Fig. 259) .... PITCHER PLANT, *Sarracenia purpurea* L.

1b Flowers yellow; leaves long, trumpet shaped, narrowly winged (Fig. 260) .... TRUMPETS, *Sarracenia flava* L.

They are difficult to keep under cultivation since bog conditions must be maintained.

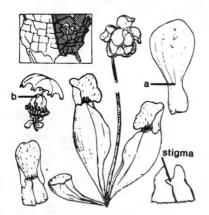

Figure 259

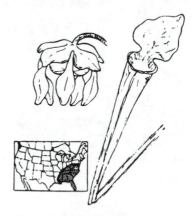

Figure 260

An insectivorous plant found in bogs in eastern North America. The hollow leaves are green to maroon, 4 to 18 inches (10-45 cm) long, and often partially filled with water and drowned insects. Because of the stiff downward pointing hairs and the waxy surface on the inside of the "pitcher," the insects cannot escape once they are inside the leaves. The plant fluids and microbes in the water of the pitchers digest the soft parts of the insects and certain of the digested chemicals are then absorbed into the plant. However, pitcher plants get most of their nutrients from the soil and through photosynthesis just as other plants do. The reddish flowers (some yellowish variants occur), 2 to 3 inches (5-7.5 cm) broad, are produced on scapes rising 8 to 20 inches (20-50 cm) tall from the basal leaves. The petals are spatulate (a), the ovary globose (b), and the broad style umbrella-shaped.

Fewer than a dozen species of these interesting plants are known from our country.

This is another spring-flowering carnivorous plant found in bogs in the southeastern corner of the United States. The flowers, 3 to 4 inches (8-10 cm) broad, are yellow, with long, drooping petals. The leaves are 1 to 3¼ feet (30-100 cm) high, yellow or yellowish green, variously veined or flushed with red or maroon. The wing is 3/8 inch (1 cm) wide or less. The roots of these plants have been used to treat indigestion and as ingredients in bitter tonics and astringents.

The insectivorous plants *Dionaea* (Venus' Flytrap) and *Drosera* (Sundew) are in another family and flower in summer except in the extreme south. Both genera grow in boggy places with the pitcher plants. The Sundews are recognized by the "dewy" glandular tentacles on the leaves; the Venus' Flytrap by the leaf with edges that snap together enclosing the prey when trigger hairs inside are touched. *Dionaea* is rare in nature, growing only in a few coastal counties of North and South Carolina. An account of all of these plants is included in *Carnivorous Plants of the United States and Canada* by Donald E. Schnell, J. F. Blair, Publishers, 1976.

CRASSULACEAE

ORPINE FAMILY

(Fig. 261) .... **WILD STONECROP,** *Sedum ternatum* Michx.

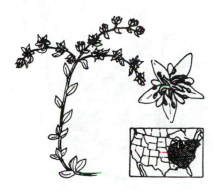

Figure 261

Found in rocky woods; also escaped from gardens. The plant, 2 to 8 inches (5-20 cm) high, has the lower leaves whorled in 3's, the upper ones alternate and sessile. The white flowers about 3/8 inch (1 cm) broad are in a leafy 2 to 4-forked cyme.

The members of the genus *Sedum,* of which something over 200 species are known, are adapted to living in desert, cold or other unfavorable conditions. They usually have fleshy stems and leaves and are often creeping in habit. They are hardy in cultivation and are often used in rock gardens.

SAXIFRAGACEAE

SAXIFRAGE FAMILY

1a Plants semi-aquatic; leaves along the stems; petals absent (Fig. 262) .... **WATER-CARPET,** *Chrysosplenium americanum* Schwein. ex Hook.

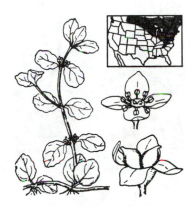

Figure 262

Shady, wet places. A low, somewhat creeping herb that forms mats in damp places. The stems are prostrate at the base, the tips ascending to 4 to 7 inches (10-18 cm) high and bearing roundish opposite leaves about 1/2 inch (1.5 cm) long. The small flowers have 4 greenish-yellow sepals and a large central disk. The salmon-red anthers (usually 8) are borne on filaments inserted in notches in the disk. The 2 styles protrude through the center of the disk. Another common name is Golden Saxifrage.

1b Plants terrestrial; leaves mostly basal; petals present 2

2a Stamens twice as many as the petals 3

2b Stamens as many as the petals (Fig. 263) .... COMMON ALUM ROOT, *Heuchera americana* L.

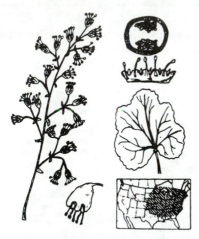

Figure 263

Dry woods. The leafless stems of this somewhat glandular-hirsute plant are 2 to 3 feet (61-91 cm) in height. The basal leaves measure 3 inches (7.5 cm) or more across and have 7 to 9 rounded lobes and long petioles. The petals are greenish to pink or purple and the stamens much longer than the remainder of the flower. The anthers are orange. The flowers are 1/4 inch (6 mm) or less in length. The Alum Root, an astringent, has been employed to relieve diarrhea and skin disorders.

A hirsute-pubscent variety of this species, *H. americana* L. var. *hispida* (Pursh) E. Wells, occurs in the same range. The flowers are irregular, with the upper parts nearly 1/16 inch (2 mm) longer than the lower. CORALBELLS, *H. sanguinea* Engelm., is a cultivated member of this genus with tall inflorescences of dainty coral pink flowers.

3a Leaves basal and in a single pair midway on the stem; petals fringed (Fig. 264) .... BISHOP'S-CAP, *Mitella diphylla* L.

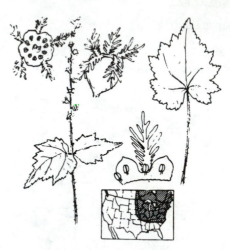

Figure 264

Rich damp woods. The flowering scape is 8 to 24 inches (20-61 cm) high with a pair of almost sessile leaves opposite each other about midway on the stem. The raceme of small dainty, white flowers has a feathery appearance because of the fringed petals. The basal leaves are ovate with a heart-shaped base and several pointed, toothed lobes. The blades and petioles are hairy. Also called Mitrewort.

NAKED BISHOP'S-CAP, *M. nuda* L., with flowering scapes having at most one small leaf and the basal leaves round to reniform with rounded lobes and crenate margins, ranges down out of the far north into New England, Michigan and Pennsylvania, and on southward on mountain tops. Also in Asia. It is a dainty little plant 2 to 8 inches (5-20 cm) high with yellowish flowers.

There are a number of similar species in the mountains of the West that have no leaves on the scapes which reach to 16 inches (40 cm) high.

3b Leaves all basal; petals entire 4

4a Flowers in a raceme (Fig. 265) .... FOAMFLOWER, *Tiarella cordifolia* L.

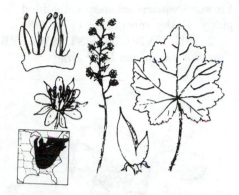

Figure 265

A delicate and beautiful plant of rich moist woods growing in southeastern Canada and northeastern United States and in the mountains to Georgia. The scape is 4 to 14 inches (10-36 cm) high with small white flowers about 1/4 inch (6 mm) broad. The feathery appearance of the flowers is due to the 10 long conspicuous stamens. The long-petioled leaves are basal. The blade is rounded, pubescent, with 3 to 5 lobes and crenate margins. The ovary is usually superior but occasionally is somewhat attached to the calyx. The capsule has 2 unequal points, giving it the appearance of a bishop's mitre. Another common name is False Mitrewort. There are also Foamflowers in the western mountains.

4b Flowers in a panicle **5**

5a Plant 1 to 3 1/4 feet (30-100 cm) high; leaves elliptic, gradually narrowed to the petiole, wavy-margined or slightly toothed; corolla greenish; petals narrow (Fig. 266) .... SWAMP SAXIFRAGE, *Saxifraga pensylvanica* L.

Figure 266

A plant of low swampy meadows and wet banks growing in southeastern Canada and in northeastern United States. The leaves, 4 to 12 inches (10-30 cm) long, are basal and from their center arises the sticky pubescent scape 12 to 39 inches (30-100 cm) long. The many small inconspicuous greenish flowers about 1/6 inch (4 mm) broad are in an open panicle. The calyx is nearly free from the ovary.

5b Plant 4 to 12 inches (10-30 cm) high; leaves spatulate, abruptly narrowed into the petiole, margins dentate or crenate; corolla white; petals broad (Fig. 267) ....
.... EARLY SAXIFRAGE, *Saxifraga virginiensis* Michx.

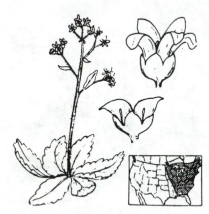

Figure 267

A common early-blooming plant of dry hill-sides and rocky woodlands. The sticky pubescent scape is 4 to 20 inches (10-51 cm) high with tiny white flowers 1/6 to 1/3 inch (4-8 mm) broad in a loose panicle. The toothed leaves, 1 to 3½ inches (2.5-9 cm) long, are basal. There are 10 stamens. The calyx, as in *Saxifraga pensylvanica*, is nearly free from the ovary. The 2-beaked seed pods are brownish-purple.

Saxifrages are flowers of cool climates. There are many species in the mountains of North America, and a number that circle the world in the higher northern latitudes. In the Pacific Northwest there are 22 species and a number of varieties.

ROSACEAE

ROSE FAMILY

1a Leaves all trifoliolate (Fig. 268) 2

1b Leaves, at least the lower, 5- to 7-foliolate or pinnately compound 4

2a Flowers white; achenes embedded in a pulpy, fleshy receptable (Fig. 268) ....
.... WILD VIRGINIA STRAWBERRY, *Fragaria virginiana* Duchesne

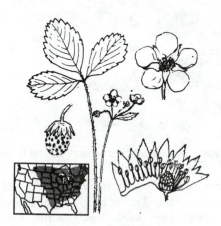

Figure 268

In dry pastures and thin woodlands. The plant is 3 to 6 inches (7.5-15 cm) high with white flowers containing numerous pistils and stamens. The leaves are compound with 3 leaflets, each leaflet toothed, the terminal tooth shorter than the ones beside it. In *F. vesca* L., another wild strawberry native in Europe as well as the eastern United States, the terminal tooth of the leaflet is longer than its neighbors. Although the wild strawberries are not as large as cultivated ones, their flavor and sweetness cannot be rivaled. They make excellent jam if one has the patience to pick them in large quantities. Bees, ants, and flies have been seen at the flowers.

The cultivated strawberry is a hybrid between *F. virginiana* and a western North and South American coastal species, *F. chiloensis* (L.) Duchesne.

2b Flowers yellow 3

3a Flowers solitary; sepals alternating with large leafy bracts; pistils numerous (Fig. 269) .... **INDIAN STRAWBERRY,** *Duchesnea indica* (Andr.) Focke

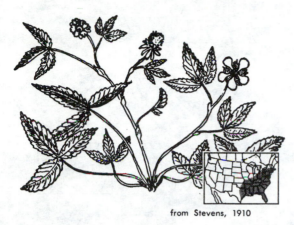

from Stevens, 1910

Figure 269

Disturbed soil, lawns and roadsides. This plant looks very much like a strawberry with extremely long creeping runners and solitary yellow flowers. The leaves are trifoliolate, the leaflets ovate or obovate and toothed or crenate. The 5 short sepals alternate with 5 large, leafy bracts, each with 3 terminal teeth. The 5 petals and numerous pistils are yellow. As the achenes mature the receptacle becomes red and fleshy, in appearance much like a strawberry, but without a sweet juicy pulp. This species is truly cosmopolitan. A native of China and India, it has been collected from North America, Mexico, Argentina and South Africa.

3b Flowers several on a scape; sepals not accompanied by large bracts; pistils 2 to 6 (Fig. 270) .... **BARREN STRAWBERRY,** *Waldsteinia fragarioides* (Michx.) Tratt.

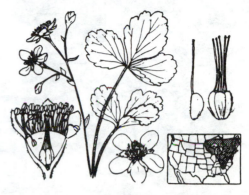

Figure 270

Wooded hillsides. The plant grows 3 to 8 inches (7.5-20 cm) high and closely resembles a strawberry except that the flowers are yellow instead of white and the leaflets are broader toward the tip and broadly wedge shaped at the base. The leaflets are irregularly toothed, and sometimes shallowly lobed as well. The bracted scapes bear 3 to 8 flowers that are 3/8 to 3/4 inch (1.0-2.0 cm) broad, with petals longer than the sepals. The fruit resembles a dried strawberry and is not edible. Another common name is Dry Strawberry.

4a Styles short; flowers yellow (genus POTENTILLA) 5

4b Styles elongate; flowers pale yellow, white or purplish (genus GEUM) 8

5a Leaves pinnately compound, leaflets 7 to 25; flowers solitary at the nodes (Fig. 271) .... SILVERWEED, *Potentilla anserina* L.

6a Flowers solitary in the axil of a leaf (Fig. 272) .... COMMON CINQUEFOIL, *Potentilla canadensis* L.

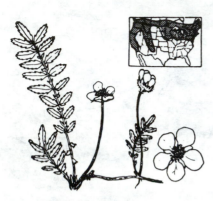

Figure 271

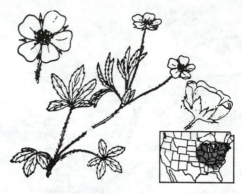

Figure 272

Moist places across the northern United States and Southern Canada and in the mountains to New Mexico and California; also in Europe and Asia. The stems are creeping, rooting at the nodes and bearing rosettes of pinnately compound leaves that are silver beneath. The leaves can reach a foot (30 cm) long; the longest leaflets, in the top 1/3, reach 1½ inches (4 cm) and become gradually shorter toward the base and the tip of the rachis. The leaflets are toothed at the margins and covered with silky pubescence beneath. The yellow flowers, 5/8 to 1 inch (1.5-2.5 cm) broad, are axillary. The fruits are achenes.

The roots are edible. When raw they have a nutty flavor; roasted or boiled, they are somewhat like parsnips.

5b Leaves palmately compound; leaflets of larger leaves 5 to 7 6

A common plant of dry soil. The plant usually grows in small rosettes, but may be somewhat creeping. The whole plant is pubescent, the undersides of the leaves densely so. The leaflets, 3/8 to 1 1/8 inches (1-3 cm) long, are obovate, wedge shaped at the base, with teeth only on the margins of the upper half. The solitary yellow flowers are 1/4 to 1/2 inch (0.7-1.3 cm) broad. There are 5 petals and 5 sepals alternating with bracts nearly the same size as the sepals.

6b Flowers many in a branched, bracteate inflorescence 7

7a Leaflets oblanceolate, toothed along the whole margin; flowers about an inch (2.5 cm) broad (Fig. 273) .... ROUGH-FRUITED CINQUEFOIL, *Potentilla recta* L.

7b Leaflets narrow, toothed beyond the middle with 2 to 4 linear teeth; flowers about 3/8 inch (1 cm) broad (Fig. 274) .... SILVERY CINQUEFOIL, *Potentilla argentea* L.

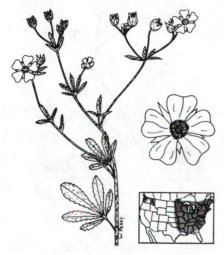

Figure 273

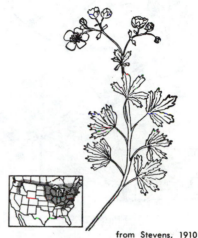

from Stevens, 1910

Figure 274

Dry soil. This is a tall plant, native of Europe. The erect stems are 16 to 32 inches (40-80 cm) tall with alternate leaves and, toward the top, several flower-bearing branches. The leaflets are usually 5, although the lower leaves may have an additional 2, smaller, leaflets. The leaflets are oblanceolate, the longest to 2 3/4 inches (7 cm) long, and deeply toothed along the whole margin. Stems, leaves, and petioles are sparsely covered with long hairs.

Naturalized in dry soil in fields and roadsides. This Eurasian native is an erect plant with stems to 20 inches (50 cm) tall. The larger leaves have 5 very narrow leaflets each with 2 to 4 oblong teeth, so that the leaves often appear rather fern-like. They are silvery underneath. The flowers have yellow petals about 1/4 inch (7 mm) long.

This genus has some 300 members which are scattered pretty much throughout the north temperate areas and even farther north. Some of them are summer flowering. The flowers are often yellow but sometimes white or purple. Several species are valued as ornamentals.

8a (4) Petals white to yellow 9

8b Petals purplish 10

9a Leaves on the lower stem 3-foliolate or only moderately dissected; petals white (Fig. 275) .... WHITE AVENS, *Geum canadense* Jacq.

9b Leaves on the lower stem highly dissected; petals yellow or cream colored (Fig. 276) .... SPRING AVENS, *Geum vernum* (Raf.) Torr. & Gray

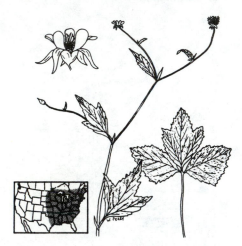

Figure 275

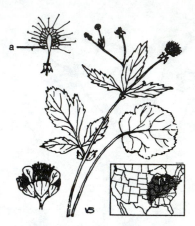

Figure 276

Woods. This is a tall plant reaching up to 3¼ feet (1 m). The basal leaves are either moderately dissected or 3-foliolate. There are a number of pointed lobes on each leaflet. The upper leaves are 3-lobed. The petals are white and about 1/4 inch (6 mm) long. The head of achenes is round.

Shady woods. This slender plant of 8 inches to 2 feet (20-50 cm) in height has several types of leaves. The basal leaves are entire or 3-foliolate with large lobes and toothed or crenate margins. The lower stem leaves are highly dissected with small, toothed lobes. The upper leaves are less dissected and with larger lobes. The small flowers have tiny yellow petals. The mature head of numerous hooked achenes is held erect on a slender stalk (a).

10a Leaves variable on a stem, the lower pinnately dissected, the upper with 3 leaflets or simply lobed, pubescent mostly on the veins (Fig. 277) .... **PURPLE AVENS**, *Geum rivale* **L.**

10b Leaves all basal, pinnately dissected, densely puberulent (Fig. 278) .... **PRAIRIE-SMOKE**, *Geum triflorum* **Pursh**

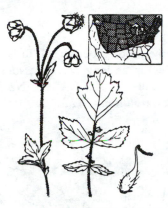

Figure 277

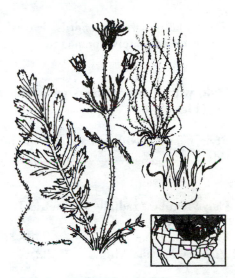

Figure 278

Bogs and wet meadows. Also in northern Europe and Asia. This plant grows 1 to 2 feet (30-60 cm) tall. The toothed leaves are variable on each pubescent stem. The lower are pinnately dissected with 3 to 5 main, more or less obovate, leaflets and a number of smaller ones mixed in. Higher on the stem the leaves are 3-foliolate or simply 3-lobed. The single flowering scape branches above and bears several nodding purplish flowers 3/4 to 1 inch (2-2.5 cm) broad. The achenes are about 3/8 inch (1 cm) long with persistent pointed styles. The roots when boiled yield a chocolaty tasting but astringent drink. With milk and sugar the drink was given as a treatment for dysentery and colic.

Dry or rocky soil. This softly pubescent plant has a rosette of oblanceolate, highly pinnately dissected leaves and one or two sets of smaller leaves on the 8 to 18 inch (20-45 cm) flowering stem. The stem branches near the top into 2 to 4, usually 3, flower stalks. The light purple flowers are up to 1 inch (2.5 cm) broad. There are triangular sepals alternating with thread-like bracts. In fruit the purplish-grey styles of the achenes are 1 to 2 inches (2.5-5 cm) long and densely plumose. From a distance the fruiting heads look like wisps of smoke. Not common.

LEGUMINOSAE [FABACEAE]

LEGUME FAMILY

1a Vining plants with tendrils; leaves pinnately compound 2

1b Erect plants without tendrils; leaves pinnately or palmately compound 6

2a Style with a tuft of hairs encircling the tip (Fig. 280-a) 3

2b Style bearded with hairs along the inner side 5

3a Calyx regular (radially symmetrical) 4

3b Calyx irregular (Fig. 279) ...... VETCH, *Vicia americana* Muhl. ex Willd.

from Stevens, 1910

Figure 279

Moist woods. The stems of these plants trail to 3¼ feet (1 m). The pinnately compound leaves have 4 to 7 pairs of ovate to elliptic leaflets with short bristles at their tips. Racemes of 2 to 9 blue-purple flowers 3/4 to 1 inch (1.5-2.7 cm) long are shorter than the associated leaves. The lower part of the calyx tube is longer than the upper.

HAIRY VETCH, *Vicia villosa* Roth, is a species introduced from Europe which also has an irregular calyx. In this vetch, the whole plant is villous, the leaflets are narrow, and the violet flowers (rarely white) are on dense racemes which are at least as long as the leaves.

4a Racemes loose, bearing 7 to 20 flowers, shorter or as long as the associated leaf (Fig. 280) .... CAROLINA VETCH, *Vicia caroliniana* Walt.

Figure 280

Growing on river banks and cliffs. A glabrous to pubescent trailing or climbing slender-stemmed plant 2 to 3¼ feet (60-100 cm) long with pinnately compound leaves. The oblong leaflets number 10 to 18 and often have mucronate tips. The flowers, 1/3 to 1/2 inch (8-12 mm) long, are 8 to 20 in a loose raceme and are whitish or pale lavender with a blue-tipped keel. The calyx is nearly regular.

4b Racemes dense, bearing 20 to 50 flowers, much longer than the associated leaf (Fig. 281) .... COW VETCH, *Vicia cracca* L.

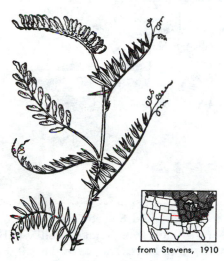

from Stevens, 1910

Figure 281

Roadsides and weedy places. The leaflets on the 20 inch to 4 foot (51-122 cm) stems are narrow, elliptic to oblong. There are several tendrils at the tips. Elongate racemes of 20 to 50 blue-violet flowers (rarely white) have long peduncles and densely crowded flowers. This species is an introduction from Eurasia.

5a Stipules broad, leaflike (a), symmetrical; plant of beaches and shores (Fig. 282) .... BEACH PEA, *Lathyrus japonicus* Willd.

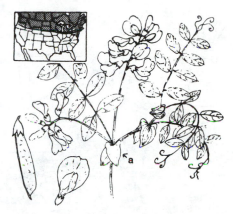

Figure 282

A stout glabrous member of the Legume Family growing on beaches from New Jersey to Arctic America, along the Great Lakes, and on the Pacific Coast. Also found in northern Europe and Asia. The pinnate leaves have 3 to 6 pairs of oval or obovate leaflets and large hastate stipules in the axils. The 5 to 10 purple flowers are 3/4 to 1 inch (1.9-2.5 cm) long. The pod averages 2 inches (5 cm) in length. The seeds contain alkaloids. They can be eaten but have not been much gathered because of the unpleasant taste.

5b Stipules narrow, assymetrical, appearing half-sagittate (a); an inland plant (Fig. 283) .... VEINY PEA, *Lathryrus venosus* Muhl. ex Willd.

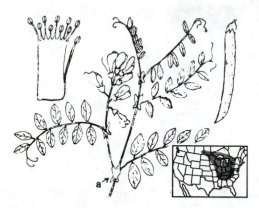

Figure 283

Shady river banks and shores. The climbing or decumbent stems are 1½ to 4 feet (46-122 cm) long with half-sagittate stipules and 8 to 14 leaflets on each leaf. The purple flowers are 1/2 to 3/4 inch (1.3-1.9 cm) long. The linear veiny pod is 1½ to 2 3/8 inches (3.8-6 cm) long.

6a (1) Leaves pinnately compound 7

6b Leaves palmately compound, leaflets 3 to 11 9

7a Racemes terminal (Fig. 284) .... WILD SWEETPEA, *Tephrosia virginiana* (L.) Pers.

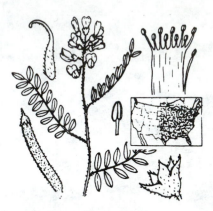

Figure 284

Dry sandy soil. Wild Sweetpea grows 8 to 28 inches (20-70 cm) high and is silky-pubescent with white hairs. The leaves bear 9 to 21 linear-oblong leaflets. The showy yellow and purple flowers 1/2 to 3/4 inch (1.3-1.9 cm) long are crowded in a terminal raceme. The densely pubescent linear pod is 1 to 2 inches (2.5-5 cm) long. Another name is Goat's Rue.

There are many species of *Tephrosia* worldwide. The roots and bark of many of these contain rotenone which has been used to stun fish to facilitate harvesting them and as a source of "organic" insecticide. *Tephrosia cinerea* (L.) Pers. is one source of the compound.

7b Racemes in the axils of leaves 8

8a Leaflets 9 to 13; flowers greenish white, less than 3/8 inch (1 cm) long; fruit an oblong pod (Fig. 285) ....
.... **LOW MILK-VETCH,**
Astragalus lotiflorus Hook.

8b Leaflets 15 to 25; flowers violet purple to whitish, 1/2 to 1 inch (1.3-2.5 cm) long; fruit a round, fleshy pod (Fig. 286)
.... **GROUND PLUM,**
Astragalus crassicarpus Nutt.

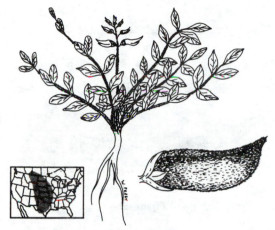

Figure 285

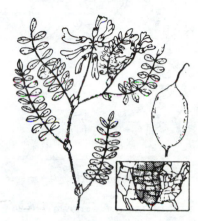

Figure 286

Sandy soil. A low plant growing in tufts. This plant has pale green leaves that are very hairy below, with 9 to 13 leaflets about a half inch (1.2 cm) long. The flowers are greenish white, 3/8 inch (8-10 mm) long, and few on a raceme. On some plants there is a second type of flower which never opens and is quite small. The pod is oblong and hairy.

The erect or decumbent stems are 8 to 20 inches (20-51 cm) long with pinnate leaves bearing 15 to 25 leaflets. The violet-purple flowers, 2/3 to 1 1/8 inch (1.7-2.8 cm) long, are in short racemes. The fruit is a glabrous oval pod with very thick walls. It is edible and is collected and stored for winter food by prairie-dogs. Another name is Buffalo-bean.
TENNESSEE MILK-VETCH, *A. tennesseensis* A. Gray, growing from Missouri and Illinois to Tennessee and Alabama differs from the above in having the pods pubescent.★

9a Inflorescence a raceme; leaves 3- to 11-foliolate **10**

9b Inflorescence a head of small flowers; leaves trifoliolate **14**

10a Leaflets 7 to 11, attached at a central point, oblanceolate; flowers usually blue (Fig. 287) .... WILD LUPINE, *Lupinus perennis* L.

Figure 287

A very beautiful and well-known 8 to 24 inches (20-61 cm) high plant of dry sandy soil. The showy purplish-blue flowers, 3/8 to 2/3 inch (1-1.6 cm) long, are in a loose raceme 4 to 8 inches (10-20 cm) in length. Sometimes the flowers are pink or white. The 4- to 6-seeded, linear-oblong pods are very pubescent.

A lupine, the TEXAS BLUEBONNET, is the state flower of Texas; it is actually the species *L. subcarinosus* rather than *Lupinus texensis*, according to a specialist in the Texan flora.

Several hundred species of lupines are known. They serve as soil builders and as ornamentals. The foliage is attractive, often covered with silvery hairs while the flowers, frequently in sizeable racemes, may be white, yellow, blue or purplish. The seeds contain the alkaloid lupinine, but if properly prepared by washing out the alkaloid, the seeds are edible and have formed part of the diet for Indians of the high Andes. In the Old World, cultivated varieties without seed alkaloids have been developed for food use.

10b Leaflets 3 to 5 .... **11**

11a Leaflets usually 5; plant densely villous; flowers bluish, in short, very dense spikes (Fig. 288) .... POMME BLANCHE, *Psoralea esculenta* Pursh

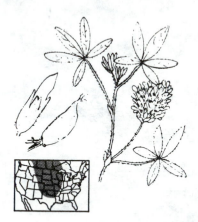

Figure 288

On prairies and plains in Manitoba and south to Texas. The stout erect plant, 4 to 16 inches (10-40 cm) high, has palmate 5-foliolate leaves and is densely villous-pubescent with white hairs. The bluish flowers, about 1/2 to 3/4 inch (1.3-1.9 cm) long, are in dense spikes. The large tuberous root contains much starch. Other names are Bread-root and Prairie Turnip.

11b Leaflets 3, obovate to oblanceolate; flowers white, cream-colored, or yellow **12**

12a Stamens grown together into 2 bunches; leaflets toothed (Fig. 289) .... MELILOTUS, *Melilotus officinalis* (L.) Pallas

13a Plant hairy; bracts of inflorescence persistent, lanceolate (Fig. 290) .... LARGE-BRACTED WILD INDIGO, *Baptisia leucophaea* Nutt.

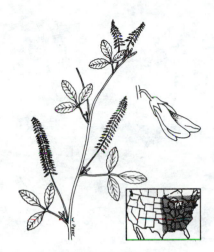

Figure 289

Figure 290

Roadsides and weedy places. This highly branched plant with stems 20 to 60 inches (50-150 cm) tall is native to Eurasia. The leaflets are oblanceolate to obovate and finely toothed on the margins. The fragrant yellow flowers are about 1/4 inch (7 mm) long in long narrow racemes.

WHITE SWEET CLOVER, *Melilotus alba* Medic., is also introduced and widespread. It has small white flowers in narrow racemes.

12b Stamens all free from each other; leaflet margins entire **13**

Prairies and woods. There are several summer-flowering Baptisias which grow to a height of 4 feet (122 cm) or more, but this April-to-June flowering species seldom exceeds 32 inches (80 cm), and branches to form clumps. The leaves are on short (less than ¼ inch [5 mm]) petioles and are villous-pubescent. The raceme is up to 8 inches (20 cm) long with the characteristic large bracts and creamy white showy flowers an inch long (2.5 cm). Livestock have been poisoned by eating this plant.

13b Plants glabrous; bracts of inflorescence deciduous (Fig. 291) .... ATLANTIC WILD INDIGO, *Baptisia lactea* (Raf.) Thieret

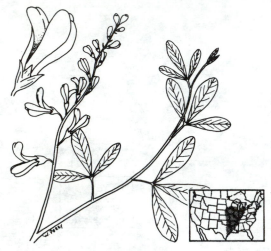

Figure 291

Synonym: Baptisia leucantha Torr. & Gray

Woods and prairies. These glabrous-stemmed plants grow to 6½ feet (2 m) tall. The leaves have petioles 1/4 to 1/2 inch (6-12 mm) long and the 3 leaflets are obovate to oblanceolate and glabrous. The white or lightly purple-tinged flowers are on racemes 8 to 24 inches (20-60 cm) tall from which the bracts fall early.

14a (9) Corolla deciduous soon after the flower opens; seed pods coiled (a) (Fig. 292) .... BLACK MEDIC, *Medicago lupulina* L.

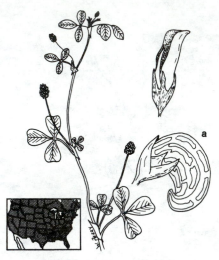

Figure 292

Weed, native to Eurasia. The plant often branches near the base into several main stems, to 32 inches (80 cm) long, which are either creeping or ascending. The leaves are clover-like, with the terminal leaflet stalked. The tiny yellow flowers are in long-stalked heads arising from the axils of the upper leaves. The pods are black and coiled into a kidney shape.

ALFALFA, *Medicago sativa* L., has escaped from cultivation and begins to bloom in spring in some areas. The flowers are up to 1/2 inch (1.2 cm) long and blue.

14b Corolla persistent, withering to brown and covering the pods; pods straight (genus TRIFOLIUM) 15

15a Flowers yellow (Fig. 293) .... LOW HOP CLOVER, *Trifolium campestre* Schreb.

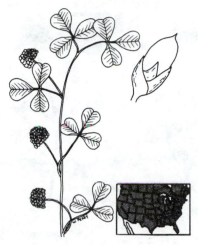

Figure 293

Synonym: *Trifolium procumbens* Schreb.

An introduced yellow clover 2½ to 16 inches (6-40 cm) high growing in sandy fields and along roadsides in much of the United States and Canada. The flower heads are 1/4 to 1/2 inch (0.6-1.3 cm) broad and become brown when dry.

YELLOW CLOVER, *Trifolium aureum* Pollick, like the above, came from Eurasia. Its terminal leaflet is sessile and the flower heads are larger.

15b Flowers white to pink or purple **16**

16a Flowers in fuzzy cylindrical heads (Fig. 294) .... RABBIT-FOOT CLOVER, *Trifolium arvense* L.

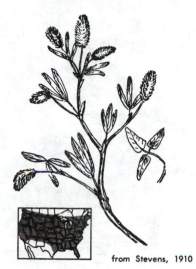

from Stevens, 1910

Figure 294

Roadsides, disturbed soil. Native in Europe and North Africa, this clover has erect-spreading, branching stems to 16 inches (40 cm) high. The leaflets are narrow and oblanceolate about ¾ inch (2 cm) long or less. The grayish ovoid to cylindric inflorescences 3/8 to 1 1/8 inches (1-3 cm) long appear furry because the densely villous-hairy calyx lobes are longer than the white to pinkish corolla.

CRIMSON CLOVER, *Trifolium incarnatum* L., is a larger clover with cylindric heads. The crimson flowers are in somewhat furry heads 1 to 2 inches (2.5-5 cm) long at the ends of pubescent stems 12 to 32 inches (30-80 cm) tall. The leaflets are broadly obovate. This species has escaped from cultivation.

16b Flowers in globose heads **17**

17a Flowers purplish red (Fig. 295) ....
.... RED CLOVER,
Trifolium pratense L.

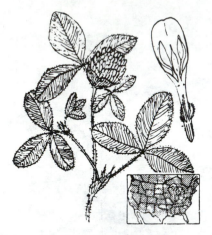

Figure 295

This very common field clover was introduced from Europe and is extensively cultivated for hay and soil enrichment. It has escaped from fields and meadows to roadsides and waste areas. The plant is 8 to 32 inches (20-80 cm) high with 3-foliolate (rarely 4 or 5) bluish-green leaves and inch-long (2.5 cm) heads of magenta flowers usually 1/2 to 3/4 inch (1.3-1.9 cm) long. This clover is almost entirely dependent upon the bumblebees for fertilization. Vermont has chosen this plant as the State Flower.

The young leaves could be used as salads or greens. Tea from the dried flower heads is considered a sedative and has been used against whooping cough.

17b Flowers white or pinkish (Fig. 296a) ....
.... WHITE CLOVER,
Trifolium repens L.

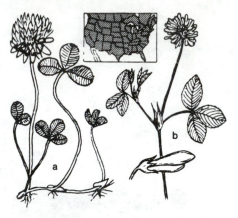

Figure 296

This is the familiar clover of lawns that yields many of the "four-leaf-clovers." It is a creeping plant which roots at the nodes. The leaves and flowers have long stalks. The leaflets are broadly obovate or elliptic, dark green, usually with an arc of white in the lower third. The flowers are white or pinkish tinged.

Too much clover causes illness in animals, but seeds and dried flowers have been made into bread during famines. This clover has made a large contribution to the cultural and aesthetic life of the United States, from songs, to greeting cards, to good luck symbols and children's clover crowns and necklaces.

ALSIKE CLOVER (Fig. 296b), *Trifolium hybridum* L., is an erect clover of meadows, waste places and roadsides. The plant is 1 to 32 inches (30-80 cm) high and has white and pink flowers.

OXALIDACEAE

WOOD SORREL FAMILY

1a Flowers rose-purple, rarely white, 1 inch (2.5 cm) broad; larger leaves 1 1/2 inches (3.8 cm) across (Fig. 297) .... VIOLET WOOD SORREL, *Oxalis violacea* L.

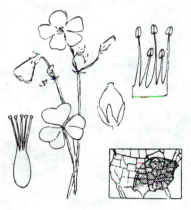

Figure 297

A small delicate plant 4 to 8 inches (10-20 cm) high found in rocky places and open woods in most of the eastern half of the United States. It is especially common in the South. The 4 to 8 leaves have 3 obcordate leaflets and are 1/2 to 1 inch (1.3-2.5 cm) wide. The scapes bear umbels of 3 to 12 rose-purple flowers 3/8 to 3/4 inch (1-2 cm) long. The leaves are all basal; they and the scapes arise from a scaly bulb-like base.

This species is a native plant but the florists also cultivate several South American and African species which have showy pink or rose blossoms.

1b Flowers yellow **2**

2a Stems erect, with spreading, blunt hairs; leaves pale green (Fig. 298) .... UPRIGHT YELLOW WOOD SORREL, *Oxalis stricta* L.

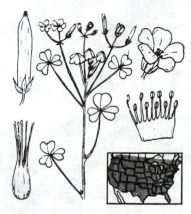

Figure 298

A common plant of roadsides, woods and fields. Although this 3 to 20 inch (7.5-51 cm) high plant is often considered a weed, it is attractive with its pale green leaves and bright yellow flowers almost 1/2 inch (1.3 cm) broad. The capsules are 1/3 to 5/8 inch (0.8-1.6 cm) long with a short beak. The fruit, leaves, and stems have a pleasant sour taste. Another name is Ladies'-sorrel.

The leaves make a wild snack, or they can be added as tangy flavoring to salads or omelets. Since the leaves contain oxalic acid, large doses are not recommended.

2b Stems creeping, rooting at the nodes; hairs sharp tipped; leaves often purplish (Fig. 299) ...... **CREEPING WOOD SORREL,** *Oxalis corniculata* L.

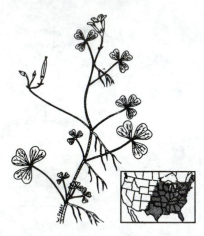

Figure 299

A tropical weed, common in the southern United States and often found in greenhouses. Leaves and flowers arise from the rooted nodes of the creeping stems. Often the trifoliolate leaves are somewhat purple tinged, the leaflets 3/8 inch (1 cm) long or less. The yellow flowers are about 1/2 inch broad (1.3 cm).

Both species of yellow-flowered Wood Sorrels are effective weeds and hard to eradicate because the ripe "pods" (capsules) open explosively at a touch and toss the small dark brown seeds a considerable distance in all directions.

GERANIACEAE

GERANIUM FAMILY

1a Leaves pinnately divided; fertile stamens 5 (Fig. 300) .... **FILAREE,** *Erodium cicutarium* (L.) L'Her.

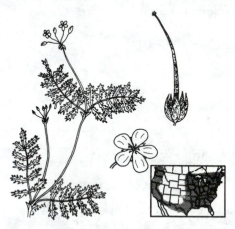

Figure 300

Fields and roadsides. The leaves of the Filaree are in a basal rosette and alternate on the flowering stems. They are long-oblanceolate in outline but pinnately divided. Each of the leaflets is pinnately toothed so the leaf looks very fern-like. The flowering stems may reach 16 inches (40 cm). Each stem bears a cluster of 2 to 8 pinkish purple flowers, about a half inch (1.3 cm) across. Other common names, Heronbill and Stork's-bill, are suggested by the long (to 1½ inches) beak of the mature ovary which protrudes from the persistent sepals.

This dainty looking plant has shown itself to be very rugged. It is native to the Mediterranean but grows in most states of the United States and has been collected as far afield as Mexico, Argentina and Chile, South Africa, and Kashmir.

1b Leaves palmately divided; fertile stamens 10 .... **2**

2a Leaves somewhat triangular in outline, divided into 3 main leaflets, each leaflet stalked, finely pinnately cut and toothed (Fig. 301) .... HERB ROBERT, *Geranium robertianum* L.

3a Leaves 2 1/4 to 6 1/4 inches wide (6-16 cm), 3- to 7-parted, the lobes broad and wedge shaped, variously cleft and toothed; flowers 1 to 1 1/2 inches (2.5-4 cm) broad (Fig. 302) ....
.... WILD GERANIUM, *Geranium maculatum* L.

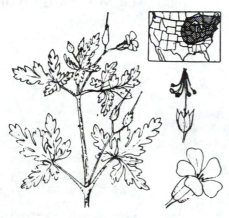

Figure 301

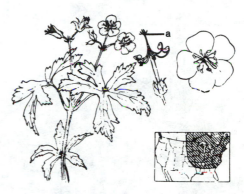

Figure 302

This species was introduced from Europe and is now established in southeastern Canada and northeastern United States. It is also found in Asia and northern Africa. It grows in rocky woods and shady areas and is 8 to 24 inches (20-61 cm) high with 3-divided, finely-cut leaves that have a strong, disagreeable odor when crushed. The reddish-purple flowers are about 1/2 inch (1-1.5 cm) broad. The beak of the fruit is 1 inch (2.5 cm) long.

2b Leaves round in outline, divided into 3 to 9 lobes which are continuous with each other near the petiole 3

An attractive and common plant of open woods and fields. The plant grows about 1 to 2 feet (30-60 cm) high and has long-petioled divided stem leaves. The rose-purple flowers, 1 to 1½ inches (2.5-3.8 cm) broad, are delicate and showy. There are 10 stamens. The beak of the fruit is 1 to 1½ inches (2.5-3.8 cm) long. When the seeds mature the carpels separate from the lower central axis of the beak while remaining attached at the tip (a).

In the eastern United States the roots were used as a source of astringent for diarrhea and sore gums and to stop bleeding. It was listed in the United States Pharmacopoeia from 1820-1916. The astringency comes from tannins.

3b Leaves 1 1/8 to 2 3/4 inches wide (3-7 cm) 5- to 9-parted, with narrow lobes, themselves often deeply and narrowly lobed; flowers pink to white, about 1/2 inch (1.3 cm) broad (Fig. 303) .... CRANESBILL, *Geranium carolinianum* L.

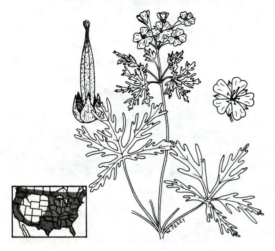

Figure 303

Dry and sandy soils. The leaves are round and finely lobed and toothed. There are leaves in a basal rosette and on the flowering stems. The stems are branched, and can be as tall as 2 feet (60 cm). The pink flowers are 4-to-many in a cluster at the tops of the stems. The beaked fruit exceeds the sepals by about 1/2 inch (1.3 cm).

POLYGALACEAE

MILKWORT FAMILY

1a Flowers 1 to 4, large, 1/2 to 3/4 inch (13-20 mm) long, rose purple (rarely white), very irregular and orchid-like (Fig. 304) .... FRINGED MILKWORT, *Polygala paucifolia* Willd.

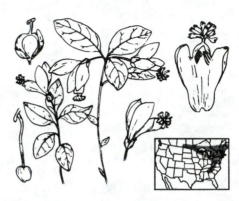

Figure 304

A dainty attractive plant of moist rich woods found from New Brunswick to Saskatchewan south to Georgia. It grows 2 to 6 inches (5-15 cm) high with 1 to 4 showy rose-purple flowers 1/2 to 3/4 inch (1.3-1.9 cm) long. Rarely the flowers are white. There are 5 sepals; 3 of them are small while the 2 lateral ones are large, colored, and petal-like. The 3 petals are united with the stamen filaments into a tube-like form, the middle or lower one with a conspicuous tufted fringe at the tip. There are 3 to 6 oval to elliptic leaves at the top of the stem. The fruit is a round capsule. Other names are Flowering Wintergreen and Gay-wings.

1b Flowers small, 1/8 inch (3-3.5 mm) long, many in a narrow terminal raceme, white or greenish white (Fig. 305) .... **SENECA SNAKEROOT,** *Polygala senega* L.

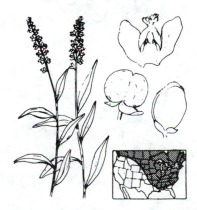

Figure 305

Rocky woodlands in much of Canada and the United States. The plant is 4 to 20 inches (10-51 cm) high with small white or greenish-white flowers in a dense raceme. The crest of the corolla is short and few-lobed. The bitter root has been used for snakebite, and diarrhea, and was listed in the United States Pharmacopoeia until 1976 as an expectorant.

RACEMED MILKWORT, *Polygala polygama* Walt., grows 4 to 20 inches (10-51 cm) high in dry sandy soil in southeastern Canada and in most of the eastern half of the United States. The showy rose-colored flowers are about 1/4 inch (6 mm) long in a loose terminal raceme.

EUPHORBIACEAE
SPURGE FAMILY

Many members of this family have milky sap, especially the tropical genera. In most of North America the main members of the family with milky sap are *Euphorbia* and *Chamaesyce*. The Poinsettia and the Castor-bean are also members of this family.

1a Bracts under the inflorescences rounded triangular, different from the stem leaves **2**

1b Bracts of the inflorescences of similar shape to the stem leaves (Fig. 306) .... **FLOWERING SPURGE,** *Euphorbia corollata* L.

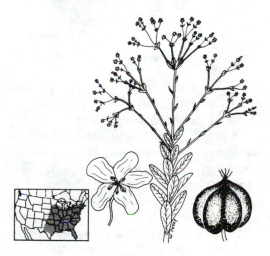

Figure 306

Dry woods, fields and prairies. The leaves of this spurge vary from obovate to lanceolate to linear. They may be as narrow as 1/8 inch (3 mm) but are usually wider. The stems, 1 to 3¼ feet (30-100 cm) tall are often several times branched in the top half. There are petal-like appendages on the 5 glands of the involucre; these are white, greenish or pinkish so that each inflorescence looks like a small flower.

2a Leaves linear, not more than 1/8 inch (3 mm) wide; involucres many on short pedicels forming a dense flat-topped terminal inflorescence (Fig. 307) .... CYPRESS SPURGE, *Euphorbia cyparissias* L.

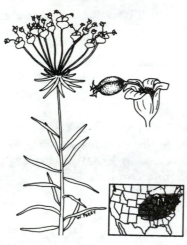

Figure 307

Cultivated and escaped to disturbed ground. The 8 to 16 inch (20-40 cm) stems of this plant have numerous narrow linear leaves, 3/8 to 1 1/8 inch (1-3 cm) long. Toward the tops of the stems there are many, bracted involucres which give the effect of a spreading flat-topped or domed inflorescence. The rounded-triangular bracts are yellowish green becoming yellowish orange with age and in comparison with the grayish-green leaves appear somewhat like flowers.

2b Leaves oblanceolate; involucres few (Fig. 308) .... TINTED SPURGE, *Euphorbia commutata* Engelm.

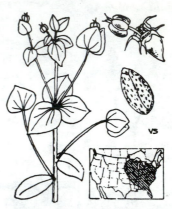

Figure 308

This somewhat-decumbent yellowish green biennial attains a height of 4 to 16 inches (10-40 cm). The leaves are often tinted with red. The umbels topping the branches are three-rayed. The glands have slender horns. The mature fruit, a capsule with rounded angles, produces seeds which are covered with pits. It is found in shady places along streams and on hillsides.

BUXACEAE

BOX FAMILY

(Fig. 309) .... **ALLEGHANY MOUNTAIN SPURGE,**
Pachysandra procumbens Michx.

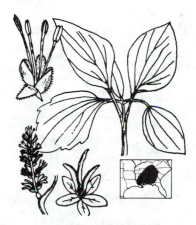

Figure 309

Woods. Thickly growing sturdy stems, a foot (30 cm) or less in height, bear broad leaves with an ovate blade 2 to 3 inches (3-7.5 cm) long and a long narrow petiole. The flower spikes arise laterally from the basal portion of the stem, below the leaves. The whitish flowers are unisexual, those with stamens are toward the top of the spike and those with pistils at the base. There are no petals and the apparent sepals are actually bracts.

PACHYSANDRA, *P. terminalis* Siebold & Zucc., with narrower but more evergreen leaves and terminal spikes is often cultivated as a ground cover and appears to be naturalized. A native of Japan, it is another of the plants which thrives in the climate of eastern North America.

LIMNANTHACEAE

FALSE MERMAID FAMILY

(Fig. 310) .... **FALSE MERMAID,**
Floerkea proserpinacoides Willd.

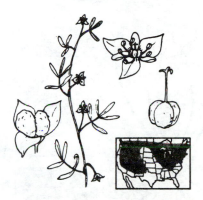

Figure 310

A rather inconspicuous plant of swampy areas in woods and river banks. The tiny white flowers have 3 sepals, 3 petals, and usually 6 stamens. This is a fragile and somewhat decumbent plant 4 to 12 inches (10-30 cm) high with pinnately compound leaves. There are 3 to 7 linear to elliptic leaflets, which are edible in green salads.

MALVACEAE

MALLOW FAMILY

(Fig. 311) .... COMMON MALLOW,
Malva neglecta Wallr.

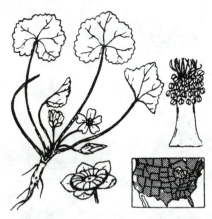

Figure 311

This small mallow is a common prostrate plant of wayside and waste places that was introduced from Europe and is now naturalized in most of North America.

The heart-shaped leaves are 3/4 to 2 3/8 inches (2-6 cm) broad on long petioles. The pinkish or bluish-white flowers are clustered among the leaves. They are 3/8 to 5/8 inch (1-1.5 cm) broad and have the stamens united into a column around the pistil. There are 12 to 15 carpels arranged in a ring, so that the fruits resemble a flat wheel of cheese, hence the common name Cheeses. The Hollyhock, Marsh Mallow, and Okra are in the same family.

Like many of the mallows, Common Mallow is mucilaginous. It has been used in medicine as a soothing agent for skin and sore throat. The young shoots and leaves are a tasty green vegetable and are particularly good as thickening agents for soups. A few young leaves floated on hot chicken broth make an interesting Chinese-style soup. The green fruit "cheeses" make a snack either fresh or pickled.

CISTACEAE

ROCKROSE FAMILY

(Fig. 312) .... FROSTWEED,
Helianthemum canadense (L.) Michx.

Figure 312

This 4 to 24 inch (10-61 cm) tall plant grows in dry sandy areas. The flowers are of two kinds: a solitary large one 3/4 to 1½ inches (2-4 cm) broad with 5 yellow petals and numerous stamens, and many smaller flowers with no petals, 3 to 10 stamens, and few seeds. The leaves are narrowly lanceolate to elliptic with stellate hairs. In the fall ice crystals sometimes shoot from the bark on the woody base.

There are six other species reported from eastern North America and a number of species in Europe.

VIOLACEAE

VIOLET FAMILY

1a Stamens united; sepals without auricles (ear-shaped basal appendage); flowers greenish white (Fig. 313) .... **GREEN VIOLET,** *Hybanthus concolor* (T. F. Forst.) Spreng.

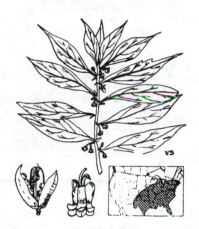

Figure 313

This erect, somewhat pubescent plant, growing in moist woods attains a height of up to 3¼ feet (1 m). The alternate leaves are 3 to 7 inches (7-17 cm) long. The small greenish white flowers about 1/4 inch (5 mm) long are borne 1 to 3 in the leaf axils. The capsule is oblong, 1/2 to 3/4 inch (1.5-2 cm) long, and at maturity splits, in typical violet fashion, into 3 valves to release large rounded seeds.

1b Stamens separate, the two lower ones spurred; sepals with auricles (ear-like lobes projecting backwards from the base, Fig. 314-a) (genus VIOLA) **2**

Figure 314

About 45 species of violets are known from the United States east of the Rocky Mountains. They are illustrated and mapped in *Violets (Viola) of Central and Eastern United States: An introductory survey* by Norman H. Russell published as an issue of the journal SIDA (Volume 2, Number 1), March, 1965.

Some of the main characters used to distinguish violets are the presence or absence of aboveground stems, the length of the protrusion of the basal petal (either a short rounded bulge or an elongate spur) and flower color (although some blue-flowered violets occasionally have white-flowered forms.) Many violet species hybridize and result in plants that show a wide variety of character combinations. In a population or over the range of the species, leaf shape and pubescence may vary depending upon the extent of hybridization with other local species or local variants of the species itself.

2a Plants stemless, the leaves and flowering scapes arising directly from a rootstock or from runners (see Fig. 315) **3**

2b Plants with leafy stems; flowers axillary **12**

3a Flowers bright yellow (Fig. 315) ....
ROUND-LEAVED YELLOW VIOLET,
Viola rotundifolia Michx.

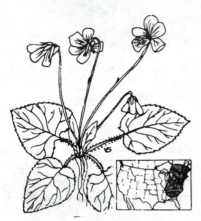

Figure 315

Cool woods, especially in beech and hemlock-deciduous forests. The leaves are oval to orbicular, on pubescent petioles. When the plant is flowering, the leaves are small, 3/4 to 1 1/8 inches (2-3 cm) long but later in summer they become as much as 4¾ inches (12 cm) long, thin, and cordate at the base. The margins are crenate. Flowering peduncles bear a single flower and are often as tall or taller than the leaves. The petals are bright yellow, the lateral ones bearded and all 3 lower ones marked with brown lines. The seeds are nearly white.

3b Flowers purple or white, or white with purple veins on the lower petals 4

4a Plants with "runners" (stolons or rhizomes); flowers 3/8 to 1/2 inch (1-1.3 cm) long, white with purple veins on the lower petals 5

4b Plants without runners; flowers 1/2 inch (1.3 cm) or longer, lilac to deep purple (occasionally with white forms) 6

5a Leaves lanceolate, tapering to the base (Fig. 316) ....
.... LANCE-LEAVED VIOLET,
Viola lanceolata L.

Figure 316

A glabrous narrow-leaved violet, 2 to 6 inches (5-15 cm) high, of moist meadows and marshes found in most of the eastern half of the United States. The fragrant flowers, 3/8 to 3/4 inch (1-2 cm) wide, are beardless and white with purple veins. The linear-lanceolate or lanceolate leaves are unusual for a violet. As in all of the violets, there are 5 stamens, 2 of which have projections which extend into the spur of the lower petal. Another name is Water Violet.

On the Atlantic and Gulf coastal plains and in Florida a variety of this species, ssp. *vittata* (Greene) Russell has linear, nearly grass-like leaves.

5b Leaves ovate to orbicular, cordate at the base (Fig. 317) .... **WILD WHITE VIOLET,** *Viola macloskeyi* Lloyd ssp. *pallens* (Banks ex DC.) Baker

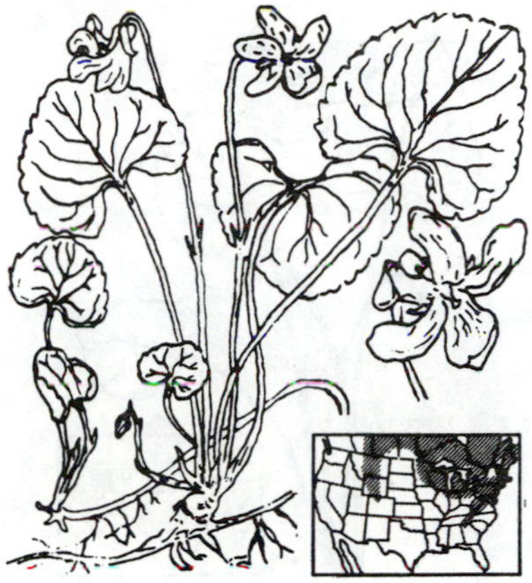

![Figure 317]

Figure 317

Synonym: *Viola pallens* (Banks) Brainerd

Wet soil. This small white violet spreads by creeping stems. The petioles of the leaves are pubescent but the ovate-orbicular leaf blades are glabrous—a combination of characters which distinguishes this species from other stemless white violets. Here the leaf blades are 5/8 to 1½ inches (1.5-4 cm) wide. The petals are white, the three lower ones marked with purple lines and the lateral ones sparsely bearded. The whole flower is about 3/8 inch (1 cm) long on a pedicel taller than the leaves. This is an Eastern and Rocky Mountain subspecies; the subspecies *macloskeyi* is found in the West.

SWEET WHITE VIOLET, *Viola blanda* Willd., occurs mostly in the Appalachians and west in the north to Ohio and southern Indiana. It is very similar to the Wild White Violet but its leaves are hairy on the upper side of the overlapping basal lobes and the tip of the leaf is acute or acuminate. Another white species, *Viola incognita* Brainerd, grows

in the same range as *V. macloskeyi* ssp. *pallens*. In this species the lower petals are definitely bearded and the leaf blades pubescent, with large spreading lobes.

6a Lateral petals beardless; flower flattened and pansy-like; leaves deeply lobed to near the base and each lobe palmately divided into narrow segments (Fig. 318) .... **BIRDFOOT VIOLET,** *Viola pedata* L.

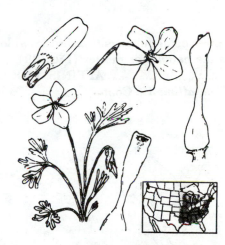

Figure 318

A large and beautiful violet growing in restricted locations in dry sandy soil on sunny hillsides or open woods. The plants are 4 to 10 inches (10-25 cm) high with leaves dissected into narrow segments, and with flattened, pansy-like flowers 3/4 to 1 3/4 inches (2-4.5 cm) wide. The petals may all be pale blue-violet or the two upper petals may be a deep velvety purple. Many communities have a "pansy hill" where these violets grow in abundance.

6b Lateral petals bearded; flowers not flattened 7

7a Leaves toothed, not lobed or parted 8

7b Leaves (at least some on each plant) palmately lobed, or parted at least on the lower half 11

8a Leaves with triangular tips, the main leaves with straight margins from mid-blade to the tip 9

8b Leaves with rounded tips, margins curved from mid-blade to tip 10

9a Lower petal pubescent; peduncles as long as the petioles or shorter (Fig. 319) .... AFFINIS VIOLET, *Viola affinis* Le Conte

9b Lower petal glabrous; peduncles longer than the petioles (Fig. 320) .... MISSOURI VIOLET, *Viola missouriensis* Greene

Figure 320

This violet of the eastern plains has glabrous leaves from a stout rhizome. The leaves are often triangular with a truncate rather than cordate base. The blades are only slightly longer than they are wide. The plant is readily recognized by its flowers which have a white center spot that is surrounded by a broad violet band on a lightly tinted violet background. The petals are about 1/2 to 3/4 inch (1.3-1.9 cm) long. The peduncles are longer than the petioles.

from Russell, 1965

Figure 319

Moist deciduous woods. This common violet is glabrous or nearly so. The larger leaves are triangular with a cordate base, up to 3 1/8 inches (8 cm) long and 3/4 to 2½ inches (2-6.5 cm) wide with toothed margins. The flower peduncles are about as long as the longest petioles but not longer. The flowers are violet-colored with a white base.

10a Lateral petals bearded with simple hairs; flowering pedicels about as long as the leaves; sepals oblong-ovate (Fig. 321) .... **COMMON BLUE VIOLET,** *Viola sororia* Willd.

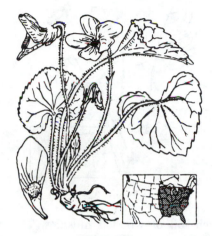

Figure 321

This is one of the most common of all of the violets, found in deciduous woods, gardens, and as a weed in lawns. The species is extremely variable. Over much of its range it is woolly pubescent and known as Woolly Blue Violet. Eastward in the range the plants become less pubescent until they are essentially glabrous and known as the Common Blue Violet. One of these glabrous plants was named *V. papilionacea*, a designation which is now considered a synonym of *V. sororia*. This violet has rounded ovate to reniform leaves, with cordate bases, which range in size from small to as large as 5 inches (13 cm) wide. The peduncles are about as tall as the leaves. The flowers may be up to 1 1/8 inches (3 cm) wide. They are usually deep purple or blue but may be lighter or even white. One variety is the Confederate Violet, a gray-white flower with blue-violet veins at the base of the petals. The capsule is mottled with brown and has dark brown seeds.

This violet is edible and healthful. The leaves, a good source of Vitamin A, can be eaten cooked, raw or wilted with a bacon and vinegar dressing or added to soup as a mucilaginous thickener. The flowers are very high in Vitamin C and make an interestingly colored salad, tea, jelly or candy. Violet syrup is a mild laxative and is used to flavor other medicines.

10b Lateral petals bearded with knob-tipped hairs; flower pedicels longer than the petioles; sepals narrowly lanceolate; growing in wet places (Fig. 322) .... **BLUE MARSH VIOLET,** *Viola obliqua* Hill

from Russell, 1965

Figure 322

Synonym: *Viola cucullata* Ait.

This is a plant of bogs or streambanks. The petioles of the ovate or reniform leaves with cordate bases are shorter than the flower stalks. The flowers are pale violet-blue with a darker center. The beards of the 2 lateral petals are dense with knobby hairs. Another name is Bog Violet.

11a Leaves elongate, at least some with toothed, sagittate bases (Fig. 323) .... **ARROW-LEAVED VIOLET,** *Viola sagittata* Ait.

11b Leaves ovate, deeply 3- to 11-lobed (lobes commonly 3 to 5) (Fig. 324) .... **EARLY BLUE VIOLET,** *Viola palmata* L.

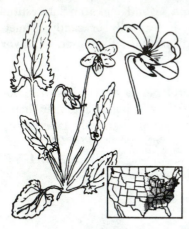

Figure 323

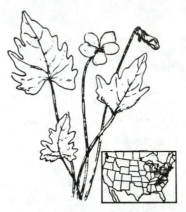

Figure 324

Moist, sunny places in sandy soil. The leaves are usually glabrous but may be pubescent. Their stems are usually longer than the blade. The blade is widened near the base into 2 backward projecting lobes which are deeply toothed in mature leaves. The flowers are deep violet-purple, up to 1 inch (2.5 cm) wide, their stems as long or longer than the leaves. It is also known as the Spade-leaf or Sand Violet.

Dry rich soil of wooded hillsides. The plant grows 3 to 8 inches (7.5-20 cm) high and has heart-shaped leaves which are palmately 5- to 11-lobed and have pubescent petioles. Sometimes the first leaves of spring are undivided. The reddish-purple flowers are 5/8 to 1½ inches (1.6-3.8 cm) broad. Like several of the blue violets, this one is sometimes called Johnny-jump-up.

The PRAIRIE VIOLET, V. *pedatifida* G. Don., of the northern Great Plains also has deeply lobed leaves but these are glabrous.

There are a number of other violet species with lobed leaves. To identify them the reader is referred to the manuals listed on page ix or the survey of Russell (see p. 149).

12a (2) Style much enlarged into a round hollow summit; stipules large, leaf-like, pectinate at the base (Fig. 325) .... **WILD PANSY,** *Viola rafinesquii* **Greene**

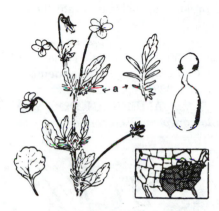

Figure 325

Synonym: *Viola kitaibeliana* R. & S.; *Viola bicolor* Pursh

Fields and open woods. The plant is 2 to 8 inches (5-20 cm) high and has small leaves 1/3 to 1/2 inch (0.8-1.2 cm) wide but large pectinate stipules (a). The small bluish-white to cream-colored flowers are not over 1/3 inch (8 mm) long. The sepals are shorter than the petals.

The common PANSY, *V. X wittrockiana* Gams, is a hybrid of *Viola tricolor* L., JOHNNY-JUMP-UP, a European native, and several other species. The style and leaves are much like the American Wild Pansy but the petals vary widely in color and size, the upper petals usually darker and more velvety than the lower. The fresh herb of *V. tricolor* is a soothing agent and was listed in the U.S. Pharmacopoeia for skin diseases.

The EUROPEAN FIELD PANSY, *V. arvensis* Murray, although similar to *V. rafinesquii*, is a sturdier plant with sepals longer than the petals. Its flowers are pale yellow with the upper petals often violet tipped. It has

escaped to cultivated fields in many places throughout the eastern half of the United States.

12b Style not enlarged into a round hollow summit; stipules ovate or lanceolate, entire or with filiform teeth but not pectinate **13**

13a Flowers yellow; leaves reniform to ovate (Fig. 326) .... **YELLOW VIOLET,** *Viola pubescens* **Ait.**

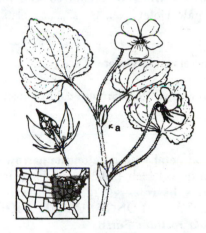

Figure 326

This species has reniform to ovate leaves with toothed margins, large stipules (a) which are entire and ovate to lanceolate, and clear yellow flowers with purple veins near the base of the petals. The two lateral petals are bearded. The species is divided into several varieties. In *V. pubescens* var. *pubescens,* DOWNY YELLOW VIOLET, there is usually one erect flowering stem 4 to 18 inches (10-45 cm) tall and 1 or 2 basal leaves. The stem leaves have 30 to 45 teeth. The plant is densely hairy and usually grows in shady, dry woods over the whole range. The SMOOTH YELLOW VIOLET, *V. pubescens* var. *eriocarpa* (Schwein.) Russell (Synonym: *V. pensylvanica* Michx.), more common in the northeastern part of the range, is glabrous or only finely pubescent and has several spreading or prostrate flowering stems up to 6 inches (15 cm) tall. There are 5 or more basal leaves. The leaves have 25 to 30 teeth. It grows in moist woods and meadows.

SPEAR-LEAVED YELLOW VIOLET, *V. hastata* is essentially an Appalachian and Lake Erie species of yellow violet, with hastate leaves. *Viola tripartita* Ell. of the southern Appalachians often has 3-parted leaves and yellow flowers.

13b Flowers white or cream colored inside, or pale violet **14**

14a Flowers pale violet **15**

14b Flowers white or cream colored inside ....
.... **16**

15a Basal petal with an elongate narrow spur 3/8 to 3/4 inch (10-20 mm) long; lateral petals beardless (Fig. 327) ....
.... **LONG-SPURRED VIOLET,** *Viola rostrata* Pursh

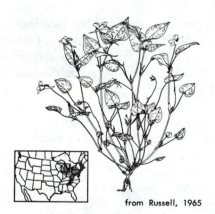

from Russell, 1965

Figure 327

Shady rich woods. This violet has many branches erect or spreading from the base.

The stem, at flowering time 2 to 5 inches (4-12.5 cm) tall, has glabrous, ovate leaves 3/4 to 1½ inches (2-4 cm) long with cordate bases and lanceolate toothed stipules with a fringe of filiform teeth. The pale violet petals have darker purple veins at their bases and the basal petal is prolonged backwards into the characteristic spur.

15b Basal petal with a rounded spur about 1/4 inch (4-5 mm) long; lateral petals bearded (Fig. 328) ....
.... **AMERICAN DOG VIOLET,** *Viola conspersa* Reichenb.

Figure 328

A plant of low, shady places, this violet flowers when the stems are only an inch tall but finally reaches to 6 to 8 inches (15-20 cm) high. The leaves have crenate-serrate margins; the lower ones are orbicular with a cordate base and the upper smaller and more nearly acute. The light violet-blue flowers usually stand well above the leaves. The lateral petals are bearded and the basal petal extends backwards between the narrowly lanceolate sepals into a rounded spur 1/8 inch or a little more (3-5 mm) long. The species is very common in the Great Lakes states and New England.

The PROSTRATE BLUE VIOLET, *Viola Walteri* House, is a relative of *V. conspersa* occurring in woods and on slopes in southern Ohio and from North Carolina and Tennessee to Florida and west to Texas. It has prostrate stems and puberulent leaves. In Canada and the northern tier of states from New England to South Dakota and in the western mountains, *Viola adunca* J. E. Smith, has dark violet flowers with short spurs and triangular leaves only slightly cordate at the base.

16a Stipules (a) oblong-lanceolate, fringed; basal petal with a spur (Fig. 329) .... **PALE VIOLET,** *Viola striata* Ait.

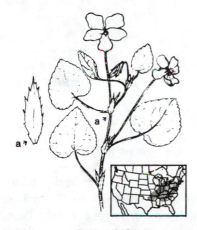

Figure 329

Found in moist woods and shady ground, this white violet is a close relative of the blue-flowered species like *V. conspersa* (Fig. 328), as can be seen from the fringed stipules and the spurred flower. The plant is 2½ to 12 inches (6-30 cm) high in the spring, but by fall may be much taller and is then inclined to be decumbent. The heart-shaped leaves have toothed edges. The flowers are whitish or cream-colored with a thick blunt spur about 1/8 inch (3-4 mm) long. Most of the violets as well as some other species of plants, have cleistogamous flowers. In the violets, cleistogamous flowers have no petals and are self-fertilized

in the bud, and only the 2 stamens that lie in the spur are developed. The capsules of the cleistogamous flowers produce many more seeds than those of the showy flowers.

16b Stipules (a) narrowly lanceolate, entire; basal petal prolonged into a short rounded sac, not markedly spurred (Fig. 330) .... **CANADA VIOLET,** *Viola canadensis* L.

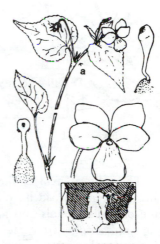

Figure 330

This is an attractive leafy-stemmed violet of woodlands. The plant is 8 to 16 inches (20-40 cm) high with cordate, toothed leaves. The fragrant flowers are almost white inside, with a yellow center, and lavender or purple outside. The lower petals are purple-veined and the two lateral petals are bearded. There are eastern and western varieties, which overlap in Wisconsin, based on leaf shape.

PASSIFLORACEAE

PASSION FLOWER FAMILY

(Fig. 331) .... **PASSION FLOWER,**
Passiflora incarnata L.

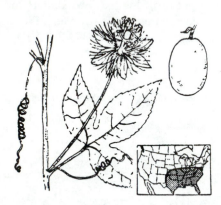

Figure 331

An attractive tendril-bearing vine 6½ to 26 feet (2-8 m) long. The solitary, axillary flowers 1½ to 2 3/8 inches (4-6 cm) broad have 5 white sepals, 5 white or bluish petals and a crown-like purple or pink fringe from the hypanthium. The 5 stamens form a tube around the stalk of the ovary which is held far above the corolla. The leaves are 2 3/8 to 6 inches (6-15 cm) broad with 3 toothed lobes. The edible yellowish fruit (a many seeded berry) about 2 inches (5 cm) long is called Maypop.

Once this plant was used medicinally as a sedative for nerves and convulsions. It was also at one time the state flower of Tennessee.

ONAGRACEAE

EVENING PRIMROSE FAMILY

1a Flowers yellow; stigma entire; leaves linear, with small marginal teeth (Fig. 332) .... **TOOTHLEAF EVENING PRIMROSE,**
Calylophus serrulatus (Nutt.) Raven

Figure 332

Synonym: *Oenothera serrulata* Nutt.

Dry prairie. This is a low bushy plant with several to many stems 8 to 22 inches high (20-55 cm) arising from a woody base. The leaves are numerous, linear, up to 2 3/4 inches (7 cm) long and about 1/4 inch (7 mm) wide, either entire or with small marginal teeth. The yellow petals are obovate, broad and up to 1/2 inch (1.3 cm) long, although often smaller. The fruits are long and narrow, in the axils of leaves. This is a very drought resistant species.

1b Flowers white to pink, with a yellow center; stigma 4-branched; leaves elliptic-lanceolate to oblanceolate with coarse teeth (Fig. 333) .... **WHITE EVENING PRIMROSE,** *Oenothera speciosa* Nutt.

Figure 333

An attractive plant of dry plains and prairies, naturalized in the South and East. The plants are 4 to 30 inches (10-75 cm) tall with flowers 1½ to 3 inches (4-7.5 cm) broad which are at first white and later turn pink. The 4 stigma lobes are linear and arranged in a cross held perpendicular to the style. The leaves are elliptic, lanceolate or oblanceolate, entire or with large teeth, or some of them pinnately divided near the base.

Another spring-flowering species is *Oenothera triloba* Nutt., THREE-LOBE EVENING PRIMROSE. This has pale yellow flowers which fade to pink, and oblanceolate, pinnately divided leaves.

The Evening Primroses open in late afternoon or at sunset. They remain open until the next morning, and rely on night-flying insects for their pollination. There are many yellow petalled summer-flowering species.

ARALIACEAE
GINSENG FAMILY

1a Leaves several, whorled midway on the stem or above, palmately compound; flowers white; fruit yellow (Fig. 334) .... **DWARF GINSENG,** *Panax trifolius* L.

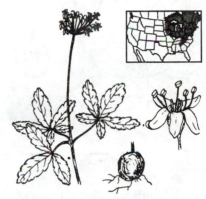

Figure 334

The Dwarf Ginseng is a small plant 2 to 8 inches (5-20 cm) high, of rich moist woods growing in southeastern Canada and in northeastern United States and in the mountains to Georgia. The whitish, often unisexual flowers, about 1/12 inch (2 mm) broad, are borne in a simple umbel. Generally there are three leaves, each with 3 to 5 leaflets. The fruit is a yellow drupe-like berry with 2 or 3 seeds. The root is globose.

GINSENG, *Panax quinquefolius* L., which resembles its dwarf cousin but blooms later, is now rare, but once was fairly abundant in woods from the Missouri River eastward. The demand in folk medicine in China and the United States for the enlarged branched roots of humanoid form of both this and the Asian species, *P. ginseng,* has led to serious over-collecting so that this species has had to be protected by law in many states.

1b Leaf 1, nearly basal, pinnately compound; flowers greenish; fruit purplish-black (Fig. 335) ...... **WILD SARSAPARILLA,** *Aralia nudicaulis* L.

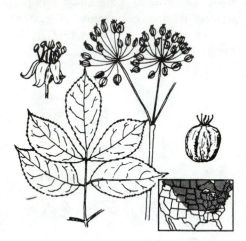

Figure 335

Moist woods. Both the single compound leaf of 3 pinnately compound leaflets up to 7 inches (18 cm) long and the flowering peduncle 3 1/8 to 8 inches (8-20 cm) long arise from a very short stem. The small greenish flowers are grouped into (usually) 3 umbels and have 5 strongly reflexed petals and 5 conspicuous stamens. The fruit is a purplish-black berry. The aromatic roots are used as a substitute for the true sarsaparilla (*Smilax*) and have been mixed by the Indians in cough remedies.

UMBELLIFERAE [APIACEAE]

PARSLEY FAMILY

The leaves in members of this family are both simple and compound. Sometimes the leaflets of the compound leaves are simple and appear very much like ordinary leaves. At other times the leaflets are themselves compound (Fig. 336).

Figure 336

The true leaf ends in a petiole which is expanded at the base and sheaths the stem. The first divisions of a compound leaf are the leaflets. The divisions of the leaflets are referred to as segments. Basal leaves may be different from the stem leaves.

The inflorescence is often a compound umbel, the rays of the primary umbel bearing smaller umbels (umbellets) at their tips (Fig. 337).

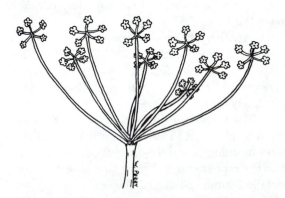

Figure 337

Although fruit characters have been avoided as much as possible here, when identifying large numbers of species in the family, fruit characters are essential.

1a Upper stem leaves divided into 3 large ovate leaflets, each leaflet 4 to 8 inches (10-20 cm) broad and palmately lobed; tall coarse herbs (Fig. 338) .... **COW PARSNIP,** *Heracleum lanatum* **Michx.**

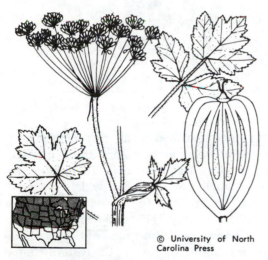

© University of North Carolina Press

Figure 338

Damp, rich soil on stream and pond banks. All parts of this coarse plant are large. The ribbed stem is 3 to 9¾ feet (1-3 m) tall and bears pubescent leaves with 3 broad, ovate leaflets that are palmately lobed. The leaf base is quite large and acts as a spathe for the young inflorescence. The umbel is compound and can be 8 inches (20 cm) broad. With its size and white flowers this is not a plant that can be overlooked.

1b Upper stem leaves with leaflets less than 4 inches (10 cm) broad **2**

2a Flowers in heads; leaves all appearing palmately compound with 3 to 7 divisions, the leaflets entire or the lateral ones 2-lobed (Fig. 339) .... **BLACK SNAKEROOT,** *Sanicula marilandica* **L.**

Figure 339

Rich woods. The plant is 1½ to 3½ feet (46-106 cm) high with 5- to 7-parted bluish-green leaves and a hollow stem. The tiny greenish-white flowers are in small clusters. The petals and sepals are about 1/16 inch (2 mm) long. The fruit is an oval bur about 1/4 inch (6 mm) long, with many hooked bristles. Another name for this common plant is Sanicle.

Sanicula canadensis L. also flowers in spring from South Dakota to Texas and east. It has white flowers, calyx lobes longer than the petals, and the bristles of the fruit longer than the styles.

2b Flowers in umbels; leaves with pinnately compound leaflets or basal leaves cordate **3**

3a Flowers white; primary rays of umbels usually less than 5 .... **4**

3b Flowers yellow; primary rays of umbels usually more than 8 .... **6**

4a Leaflet segments ovate to lanceolate, toothed, 3/8 to 1 1/2 inches (1-4 cm) wide (Fig. 340) .... ANISE ROOT, *Osmorhiza longistylis* (Torr.) DC.

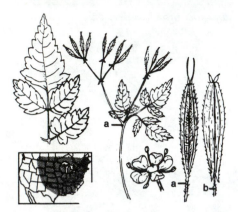

Figure 340

Rich woods. This plant, 1½ to 3 feet (40-91 cm) high, has compound, toothed leaves with ovate to lanceolate segments 3/8 to 1½ inch (1-4 cm) wide. The few-rayed umbel carries umbellets with 3 to 6 perfect, whitish flowers and some staminate flowers. The fruit is about 1/2 inch (1.3 cm) long with long, persistent styles (a) and has a pleasant aromatic odor when crushed. The aromatic root has been used to relieve digestive gas pains, but since these plants are very similar to Poison Hemlock (*Conium maculatum* L.), the drink that executed Socrates, it is suggested that this plant not be used as a home remedy.

WOOLLY SWEET CICELY, *O. claytonii* (Michx.) C. B. Clarke, is much like *O. longistylis* in the same range except that the styles are shorter in the fruit (b) and the foliage is not anise scented but is more woolly pubescent.

4b **Leaflets finely divided into narrow segments, the segments themselves also deeply lobed or divided** **5**

5a Fruits narrow, several times longer than wide; plants 8 to 24 inches (20-60 cm) tall (Fig. 341) .... SPREADING CHERVIL, *Chaerophyllum procumbens* (L.) Crantz

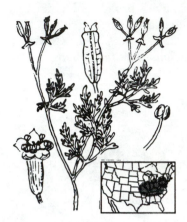

Figure 341

This much-branched, weak, sprawling plant is found on moist ground and attains a height of 8 to 24 inches (20-61 cm). The white flowers are small, few and rather widely scattered. The leaves are very finely divided; the smallest divisions only 1/16 inch (2 mm) wide.

Several similar members of the genus, native or introduced, grow in the eastern United States and Canada.

5b Fruits less than twice as long as wide; plants 2 to 6 inches (5-15 cm) tall at flowering, taller in fruit (Fig. 342) .... **HARBINGER-OF-SPRING,** *Erigenia bulbosa* (Michx.) Nutt.

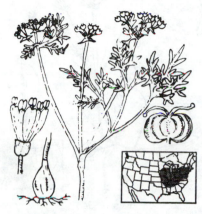

Figure 342

This low-growing glabrous perennial rises to 2 to 6 inches (5-15 cm) from a deep tuber at flowering time. The stem continues to grow as the fruits mature. The leaves are finely cut and much divided, the final divisions less than 1/4 inch (6 mm) broad. The small white petals are flat and the fruit wider than long. The white petals and dark anthers give this plant a salt-and-pepper appearance as it blooms very early in the spring. It is visited by early-flying small insects. This is the only species in the genus.

6a Basal leaves cordate (Fig. 343) .... **GOLDEN PARSNIP,** *Zizia aptera* (Gray) Fern.

Figure 343

This sturdy plant of open woods and prairies grows to a height of 2 to 3 feet (60-90 cm). It has long-petioled cordate basal leaves usually 2 inches (5 cm) wide but occasionally wider. The stem leaves are compound. The flowers are yellow and mature into a fruit with 10 ribs.

6b Basal leaves not cordate 7

7a Leaflets finely divided into narrow segments, the segments themselves also lobed or divided (Fig. 344) .... PRAIRIE PARSLEY, *Polytaenia nuttallii* DC.

8a Final leaf divisions entire, obovate to oblong (Fig. 345) .... GOLDEN ALEXANDER, *Taenidia integerrima* (L.) Drude

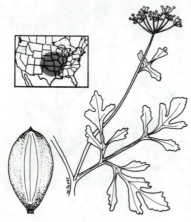

Figure 344

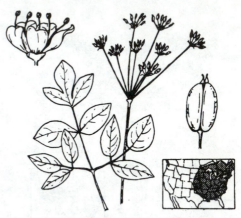

Figure 345

Prairies. From April to June the plants bloom with yellow flowers on stems 20 to 40 inches (50-100 cm) tall. The leaves are large and several times pinnately divided; the final divisions have a number of teeth. The fruit is oval, about 3/8 inch (1 cm) long, and surrounded by a thick wing.

Rocky or sandy soil. The plant is 8 to 32 inches (20-80 cm) high with small yellow flowers in compound umbels. The rays of the primary umbels are of different lengths. The fruits are oval and about 1/8 inch (3-4 mm) long. The leaves are 2 to 3 times compound; the final divisions are obovate to elliptic or oblong with entire margins. The whole plant is glabrous. Another name is Yellow Pimpernel.

7b Leaflet segments entire or toothed but not deeply lobed or divided 8

8b Final leaf divisions toothed 9

9a Margins of leaf segments deeply and irregularly toothed (Fig. 346) ....
.... **HAIRY-JOINTED MEADOW**
                                    PARSNIP,
Thaspium barbinode (Michx.) Nutt.

9b Margins of leaf segments regularly toothed (Fig. 347) ....
.... **EARLY MEADOW PARSNIP,**
Zizia aurea (L.) W.D.J. Koch

Figure 346

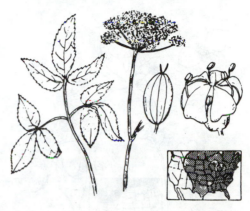

Figure 347

This erect plant grows along streams and attains a height of 2 to 3¼ feet (60-100 cm). The stems are heavily pubescent at the nodes; the leaves mostly bi-pinnate (twice compound). The coarsely and irregularly toothed leaf segments are ovate to lanceolate and up to 1 3/8 inches (3.5 cm) wide. The flowers are yellow on compound umbels. The fruit is elliptical, about 1/4 inch (6 cm) long and bears wings, as shown in the cross section drawing (a).

A common plant, 1 to 2½ feet (30-76 cm) tall, of fields and roadsides. The tiny yellow flowers are in many small clusters in a compound umbel with 8 to 18 primary rays. The leaf segments are ovate to lanceolate, 3/4 to 2¼ inch (2-6 cm) long and 3/8 to 1 1/8 inch (1-3 cm) wide with regularly toothed margins. The fruit is oblong. The plant begins to bloom in April.

CORNACEAE

DOGWOOD FAMILY

(Fig. 348) .... BUNCHBERRY, *Cornus canadensis* L.

Figure 348

Moist woods. A striking plant 4 to 8 inches (10-30 cm) high. What seems to be a single flower is really a head of many tiny greenish flowers surrounded by 4 to 6 white petal-like bracts. Each true flower has 4 stamens and 4 petals, one of the petals with an appendage. The head of fruit is composed of brilliant red drupes. Four to six leaves are in a whorl near the top of the single stem. The leaves are obovate, to oblanceolate or lanceolate and pointed at both ends. As in the other Dogwoods the veins of the leaves arc forward and parallel the leaf margins. This species also grows in eastern Asia. This is the herbaceous sister of the Flowering Dogwood tree, *Cornus florida* L.

ERICACEAE

HEATH FAMILY

1a Calyx and corolla 4-parted or lobed; stamens 8; plant creeping, slightly woody; fruit a white berry (Fig. 349) .... CREEPING SNOWBERRY, *Gaultheria hispidula* (L.) Muhl. ex Bigelow

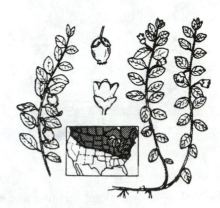

Figure 349

In cold woods and bogs. This species is a dainty creeping plant with small stiff evergreen leaves 3/8 inch (1 cm) long and tiny white bell-shaped flowers. The leaves are pubescent with brownish hairs beneath. The 4-lobed nodding flowers, about 1/6 inch (4 mm) long, grow in the leaf axils and are followed by pure white, aromatic edible berries. Both the fruit and the leaves have a wintergreen flavor. Another name is Moxie Plum.

The WINTERGREEN, *G. procumbens* L., may be in bloom in early spring in woods on acid soil. This species has erect stems bearing elliptic to ovate evergreen leaves up to 2 inches (5 cm) long. The flower parts are in 5's, the petals white or pink, and the berry is red.

1b Calyx and corolla 5-parted or lobed; stamens 10; plant trailing, woody (Fig. 350) .... TRAILING ARBUTUS, *Epigaea repens* L.

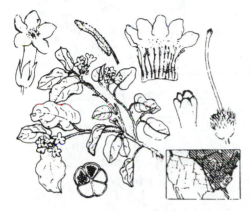

Figure 350

Sandy woods and rocky soil. Probably one of the most celebrated of all spring flowers, as it well deserves to be. The tough, shrubby stems creep over the ground often forming large patches. The thick oval or orbicular leaves, 3/4 to 3½ inches (2-9 cm) long, and the stems are covered with rust-colored stiff hairs. The white or pinkish blossoms are in few-to-several-flowered clusters and have a rich spicy fragrance. The corolla tube is 1/4 to 5/8 inches (0.6-1.5 cm) long; there are 5 locules in the ovary and 5 lobes in the stigma. This flower has disappeared from many areas because it was so extensively gathered. It is now frequently protected from picking. Another name is Mayflower.

PRIMULACEAE

PRIMROSE FAMILY

1a Leaves basal; flowers in an umbel at the top of a leafless scape 2

1b Leaves on the stem; flowers terminal or axillary 3

2a Calyx and corolla reflexed; the corolla longer than the calyx; leaves 2 1/4 to 8 inches (6-20 cm) long (Fig. 351) .... SHOOTING-STAR, *Dodecatheon meadia* L.

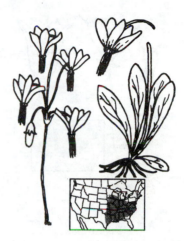

Figure 351

An attractive plant of moist cliffs and prairies. The oblong or oblanceolate leaves are basal and from their center arises a scape 8 to 24 inches (20-61 cm) high, which bears purple, pink or white flowers 3/4 to 1 3/8 inches (1.7-3.5 cm) long, in an umbel. The corolla lobes are strongly turned back exposing the 5 stamens which are close together in a cone-shaped form. Both the filaments and the corolla base are marked with dark purple. Another name is American Cowslip.

Fifteen species of Shooting-star grow in North America, most of them in the western mountains and foothills.

The cultivated Cyclamen, which also has reflexed petals, is a member of this family.

2b Calyx and corolla erect, corolla shorter than the calyx; leaves less than 5/8 inch (1.5 cm) long (Fig. 352) .... ANDROSACE, *Androsace occidentalis* Pursh

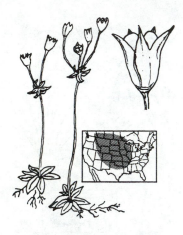

Figure 352

This tiny plant growing from 1 to 3 inches (2-7.5 cm) high lives in dry soil. The oblong or spatulate leaves are about 1/2 inch (6-15 mm) long and are usually covered with a minute pubescence. One to several scapes, 3/4 to 2 inches (2-5 cm) tall from base to the bottom of the umbel, may arise from each rosette. The flowers are on individual pedicels up to 1 1/8 inch (3 cm) long from a whorl of bracts. The corolla is white but mostly hidden by the pale tubular calyx.

Species of *Androsace* are characteristic of arctic and alpine regions around the world. Three other species are native in North America and there are Androsaces in the Alps, the Pyrenees, the Himalayas, the Pamirs and more. Because the rosettes of leaves hug the ground they are warmed by heat radiated from the soil and rock, and protected from cold winds so that they are able to get an early start on the short growing season in cold climates.

3a Leaves in one whorl at the summit of the stem; flowers white or pink (Fig. 353) .... STARFLOWER, *Trientalis borealis* Raf.

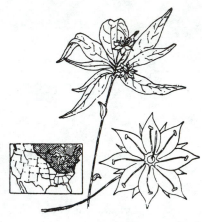

Figure 353

A dainty, fragile woodland plant found in much of the eastern half of Canada and the United States. It grows 3 to 9 inches (7.5-23 cm) high with the 5 to 10 leaves in a whorl at the summit. Usually there are 2 white (rarely pink) sharp-petaled flowers, 1/3 to 1/2 inch (8-13 mm) broad, in the center of the leaves. The corolla lobes number 5 to 9, but are usually 7. It has a horizontal root stalk which may grow either on or under the ground.

The subspecies *T. borealis* ssp. *latifolia* (Hook.) Hultén with pinkish flowers occurs in the western fourth of North America.

3b Leaves mostly opposite (the species mentioned incidentally is whorled); flowers yellow, salmon-red or blue, rarely white **4**

4a Flowers salmon-red or blue (rarely white), to 5/8 inch (1.5 cm) broad, solitary in leaf axils (Fig. 354) .... **PIMPERNEL,** *Anagallis arvensis* **L.**

Figure 354

A plant of waste places that was introduced from Europe and is now established in most of the eastern half of North America and along the Pacific coast. It is usually much branched with each branch 4 to 12 inches (10-30 cm) long bearing small flowers, 1/4 to 1/2 inch (6-13 mm) broad, in the leaf axils. The ovate or oval leaves are usually opposite but occasionally are in 3's.

The common variety, var. *arvensis,* SCARLET PIMPERNEL, has scarlet or salmon colored flowers which make it an attractive weed. The more rare var. *coerulea* (Schreb.) Gren. & Godr. has blue flowers. The species has been reported to be both poisonous and edible, as well as a remedy for both dropsy and toothache!

4b Flowers yellow, small and clustered in the leaf axils or 3/4 to 1 1/4 inch (2-3 cm) broad and solitary **5**

5a Plant creeping; flowers 3/4 to 1 1/4 inch (2-3 cm) broad (Fig. 355) .... **MONEYWORT,** *Lysimachia nummularia* **L.**

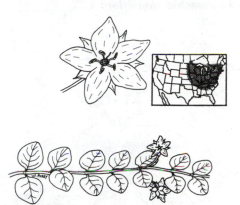

Figure 355

Moist places. The creeping stems have opposite leaves that are oval to nearly round, about 1 3/8 inch long (3.5 cm) or smaller. The pedicels of the bright yellow flowers are somewhat longer than the leaves. The petals are dotted with dark red. This is a European native which has escaped from cultivation.

In late May, *L. quadrifolia* L., an upright species with whorled leaves begins to bloom in open woods from Maine to Wisconsin to Alabama and South Carolina. The yellow flowers (about 3/4 inch [2 cm] broad) are solitary in the axils of the upper leaves.

5b Plant erect; flowers less than 1/2 inch (1 cm) broad, in dense axillary racemes (Fig. 356) .... **TUFTED LOOSESTRIFE,** *Lysimachia thyrsiflora* **L.**

LOGANIACEAE

LOGANIA FAMILY

(Fig. 357) .... **INDIAN PINK,** *Spigelia marilandica* **L.**

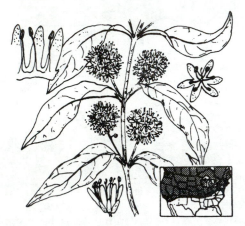

Figure 356

Figure 357

A plant of swamps, growing from Nova Scotia to Alaska and California. Also found in Europe and Asia. The simple erect stems are 1 to 2½ feet (30-76 cm) high with opposite lanceolate leaves and dense racemes of yellow flowers in the leaf axils. The flowers, about 3/8 inch (1 cm) broad, are spotted with black. This plant blooms from May to July.

The Indian Pink is a showy plant 1 to 2 feet (30-61 cm) high which grows in woods. Usually the flowers are in a single terminal one-sided spike. The corolla is scarlet outside, yellow inside, 1 to 2 inches (2.5-5 cm) long. The calyx is 5-parted, the divisions pointed, linear. The sessile ovate-lanceolate leaves are opposite. Other names are Carolina Pink, Wormgrass and Pink-root. The root was used by Indians and listed in the official Pharmacopoeia from 1820 to the early 1900's as a remedy for intestinal worms. This species is the northernmost member of a genus that is essentially Central and South American.

GENTIANACEAE

GENTIAN FAMILY

(Fig. 358) .... **PENNYWORT,**
Obolaria virginica L.

MENYANTHACEAE

BUCKBEAN FAMILY

(Fig. 359) .... **BUCKBEAN,**
Menyanthes trifoliata L.

Figure 358

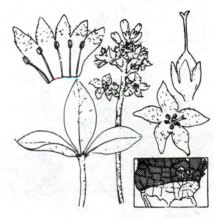

Figure 359

This purplish-green plant 1 1/8 to 6 inches (3-15 cm) high grows in moist thickets and rich woods. The lower stem has several pairs of short, scale-like leaves around the flowers and several purplish leaves 1/4 to 1/2 inch (5-15 mm) long. The flowers are sessile and have 2 sepals, 4 corolla lobes and 4 stamens. The flowers are white, purplish or pink.

The familiar blue Gentians which give the family its name are summer and fall flowering plants. In the south, a few pink- or white-flowered species of *Sabatia*, MARSH PINK, with short corolla tubes and spreading lobes, begin to flower in late spring.

An attractive plant, 4 to 15 inches (10-38 cm) high, of bogs and shallow water found in Alaska, most of Canada and in the United States. This plant has a thick rootstock from which arise stout 3-foliolate leaves and a raceme of 10 to 20 flowers. The purplish or white corolla is bearded with white hairs and is about 1/2 inch (1.3 cm) broad. This blooms from May to July and is also called Marsh Trefoil.

The leaves and rootstocks contain a bitter substance that is used in tonics to stimulate digestion and has also been used as a substitute for hops in brewing.

APOCYNACEAE

DOGBANE FAMILY

1a Plant trailing; flowers solitary in the leaf axils (Fig. 360) .... PERIWINKLE, *Vinca minor* L.

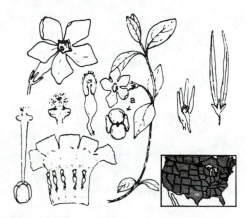

Figure 360

A very attractive trailing plant that was introduced from Europe and has escaped from cultivation to roadsides and woods mostly in the Eastern States. The firm, dark green, shiny leaves are opposite and bear the purplish-blue flowers 5/8 to 1 1/8 inches (1.6-2.8 cm) broad in their axils. There are 2 distinct carpels which alternate with the 2 glands of the disk. The fruit consists of 2 narrow follicles. Another name is Myrtle.

1b **Plant erect; flowers in cymes at the ends of branches** 2

2a Leaves alternate; flowers blue with long, spreading corolla lobes (Fig. 361) .... AMSONIA, *Amsonia tabernaemontana* Walt.

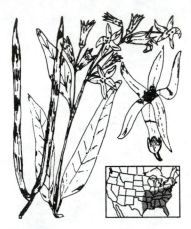

Figure 361

Moist soil. This interesting plant with lanceolate to ovate leaves 2½ to 6 inches (6-15 cm) long is mostly glabrous and sparingly branched. The inflorescences are loose and pyramid shaped. The blue flowers have a small calyx and a pubescent corolla tube 3/8 inches (1 cm) long and 5 elongate lobes which spread, star-like, to nearly 1 inch (2.5 cm) across. The fruit consists of twin follicles 3 to 4¾ inches (7.5-12 cm) long which usually stand erect. Another name is Blue-star.

2b Leaves opposite; flowers white with short corolla lobes (Fig. 362) .... **INDIAN HEMP,** *Apocynum cannabinum* L.

Figure 362

Sunny places. Opposite leaves on petioles about 1/4 inch (6 mm) long help to distinguish this plant. The stems are more than 3 feet (90 cm) tall and often end in a more or less flat-topped inflorescence of small white or greenish-white flowers. The flowers are cylindrical or urn shaped, about 1/4 inch (6 mm) long and less than that wide. The leaves are elliptic to lanceolate with a mucronate tip. These plants poison grazing cattle and sheep.

ASCLEPIADACEAE

MILKWEED FAMILY

In the genus *Asclepias,* to which all of the plants keyed here belong, the flower parts are modified and often elaborate (Fig. 363).

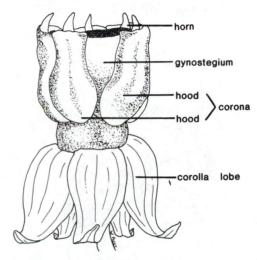

Figure 363

There are 5 small sepals and 5 corolla lobes, which are reflexed in most species. There is also a corona of 5 fleshy hoods, each hood often with a small horn projecting from the center. There are 2 ovaries but one stigma. The anthers adhere to the stigma to form a central gynostegium.

1a Leaves thin, petiolate at 3 (or 4) nodes on the stem, opposite at the upper and lower nodes and in whorls of 4 at the middle node (if 4 nodes, opposite at the middle nodes), ovate to lanceolate; leaf tips acuminate (Fig. 364) .... **FOUR-LEAVED MILKWEED,** *Asclepias quadrifolia* Jacq.

Figure 364

Found in dry woods and hills. The slender plant, 8 to 20 inches (20-50 cm) high, has thin ovate or lanceolate leaves. Some of them are in whorls of 4. The largest leaves are 4¾ inches (12 cm) long. There are 1 to 4 usually terminal umbels of pink or white flowers. The corolla lobes are not over 1/4 inch (6 mm) long. The hoods of the corona are taller than the gynostegium and there is a distinct curved horn shorter than the hood (a). Fruits are long and narrow.

The WHORLED MILKWEED, *A. verticillata* L., has narrow linear leaves in many whorls on the stem. It ranges from Massachusetts to Florida and Saskatchewan to Arizona, but begins to bloom by May only in the southern part of its range.

1b Leaves thick, usually at 4 or more nodes, opposite, sub-opposite or alternate; leaf tips acute, obtuse or rounded **2**

2a Petals reflexed; horns present **3**

2b Petals not reflexed; horns absent (Fig. 365) .... **GREEN ANTELOPE-HORN,** *Asclepias viridis* Walt.

Figure 365

Dry woods and prairies. This 2 foot (60 cm) tall milkweed has mostly alternate leaves which are elliptic or oblong, with rounded or obtuse tips. The large flowers have greenish petals 3/8 to 5/8 inch (1.0-1.7 cm) long which spread upwards around the rest of the flowers. The hoods are purple, about the same height as the gynostegium, and have no horns. Fruits are spindle shaped.

Fruits of all the milkweeds are follicles which open to release flat brownish seeds, each with a tuft of long white hairs that help to disperse the seed on gusts of wind.

3a Leaves sessile, with ruffled margins; horns longer than the hoods (Fig. 366) .... BLUNTLEAF MILKWEED, *Asclepias amplexicaulis* Sm.

3b Leaves petiolate, without ruffled edges; horns shorter than the hoods (Fig. 367) .... COMMON MILKWEED, *Asclepias syriaca* L.

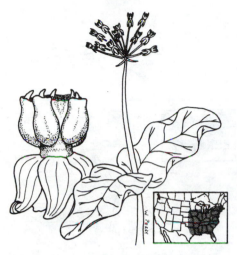

Figure 366

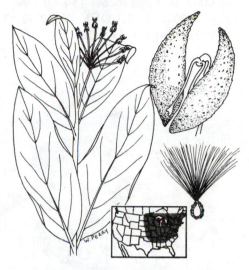

Figure 367

Dry areas, often in sandy soil. The 39 inch (1 m) stems produce 2 to 6 pairs of opposite, oval or oblong leaves with ruffled margins. The bases are cordate and, being sessile, appear to clasp the stem. The single, terminal inflorescence has many flowers with reflexed purple corolla lobes about 3/8 inch (1 cm) long. The horns are longer than the hoods and arch toward the center of the flowers.

Although very young shoots and seedpods of *Asclepias* are edible boiled as an asparagus-like vegetable, great care must be used to distinguish them from the Dogbanes, *Apocynum,* which are poisonous. Later the stems and foliage of Milkweeds develop poisonous cardiac glycosides which cause severe digestive upset and eventual heart disturbances in humans and other animals. A few caterpillars live by chewing the leaves, but these caterpillars (and the butterflies, such as the Monarch, they become) store the plant's glycoside and poison birds which are unwary enough to eat them.

Terminal and axillary inflorescences and petiolate leaves up to 10 inches (25 cm) long and 4 inches (10 cm) wide and on stems 1 2/3 to 6½ feet (0.5-2 m) tall. The leaves are flat, thick, and elliptic to oblong with acute or mucronate tips. The rose or greenish-white corolla lobes are reflexed. The hoods are taller than the gynostegium and the horns are shorter than the hoods.

The BUTTERFLYWEED, *Asclepias tuberosa* L., has red to yellow flowers with reflexed petals and many lanceolate to linear leaves scattered on the stem. It is a summer-flowering species that begins to bloom in May. It grows over most of the eastern United States.

CONVOLVULACEAE

MORNING-GLORY FAMILY

1a Calyx hidden by 2 large bracts attached to the pedicels close under the flower; corolla 2 to 2 3/4 inches (5-7 cm) broad (Fig. 368) .... **HEDGE BINDWEED,** *Calystegia sepium* (L.) R. Br.

1b Calyx not hidden by bracts, small bracts about midway on the pedicel; corolla 1 inch (2.5 cm) broad (Fig. 369) .... **BINDWEED,** *Convolvulus arvensis* L.

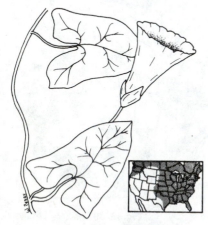

Figure 368

Synonym: *Convolvulus sepium* L.

Disturbed soil. A twining or trailing vine 1 2/3 to 9¾ feet (0.5-3 m) long, this plant has triangular leaves 2 to 4 inches (5-10 cm) long with sagittate bases and long petioles. The flowers stand above the leaves, solitary, on long peduncles. The calyx is nearly hidden by the 2 ovate bracts. The corolla is rose-purple or white, funnel shaped, with no free lobes but 5 scarcely-defined angles on the limb. The plant is native to temperate regions of North America and Eurasia.

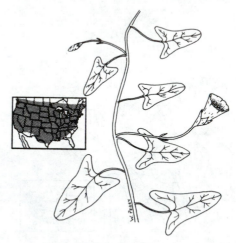

Figure 369

Disturbed sites, fields and gardens. This Bindweed is essentially a smaller version of the previous species. The stems trail to 3¼ feet (1 m). The leaves are variable, broadly triangular or narrow, pointed or triangular at the tip, the bases hastate or sagittate, sometimes only 1/2 inch (1.3 cm) long but up to 2 inches (5 cm) in length. The flowers are white or pinkish, sometimes with rosy streaks on the outside. It would be an attractive garden plant if it could be contained. The rhizomes spread underground and if broken by spading, simply produce more plants from each small piece, which then spread by their own rhizomes. Enough nutrients are made by one part of the plant to support tens of feet of whitened vine growing in dark sheds or under and through loose piles of wood.

POLEMONIACEAE

POLEMONIUM FAMILY

1a Leaves alternate, pinnately compound; leaflets ovate-oblong; flowers campanulate (Fig. 370) .... **GREEK VALERIAN,** *Polemonium reptans* **L.**

Figure 370

Woodlands. The glabrous plant is 8 to 16 inches (20-40 cm) high with weak slender stems and pinnately compound leaves. The sky-blue flowers, 3/8 to 2/3 inch (1-1.7 cm) broad, are in loose panicles. The stamens are as long as the corolla or shorter. Another name is Jacob's-Ladder. ★ — TX

AMERICAN JACOB'S-LADDER, *Polemonium van-bruntiae* Britton, is a similar plant of swamps and stream banks growing in Vermont and New York to Maryland. The erect plant is 1 1/3 to 3¼ feet (40-100 cm) high with bluish-purple flowers 2/3 to 3/4 inch (1.7-2 cm) broad. In this species the stamens are longer than the corolla.

1b Leaves opposite or whorled, simple; flowers with a narrow tube and nearly perpendicular spreading (salverform) lobes

.... **2**

2a Leaves subulate (awl-shaped) (a), some of them whorled or crowded; plant decumbent (Fig. 371) .... **GROUND PINK,** *Phlox subulata* **L.**

Figure 371

A common flower of dry sandy soil and rocky hills, and often cultivated. The tufted stems are branched and form green mats sprinkled with the attractive pink, purple, or white flowers. The corolla has a dark center and is 1/2 to 3/4 inch (1.3-2.0 cm) wide with notched obovate lobes. Another name is Moss Pink. This species blooms from April to June.

Four other similar species live on dry hills and mountains of the East and West.

2b Leaves ovate, lanceolate, or linear, opposite **3**

3a Flowers in cymes in a narrow panicle; stem flecked with purple (Fig. 372) .... **WILD SWEET WILLIAM,** *Phlox maculata* L.

Figure 372

Rich woodlands. The slender stem is 1½ to 3 feet high and is usually beautifully flecked with purple. The leaves are lanceolate or ovate-lanceolate, glabrous and firm. The pink or purple flowers are generally in a long narrow panicle. The corolla tube is 3/4 to 1 inch (1.9-2.5 cm) long and the corolla limb about 7/8 inch (2.2 cm) wide.

3b Flowers scattered or in corymbose cymes **4**

4a Leaves linear or linear-lanceolate (Fig. 373) **7**

Figure 373

4b Leaves ovate or ovate-lanceolate (Fig. 374) **5**

Figure 374

5a Plants pubescent or hirsute **6**

5b Plants glabrous or almost wholly so; calyx teeth acute (Fig. 375) .... **MOUNTAIN PHLOX,** *Phlox ovata* L.

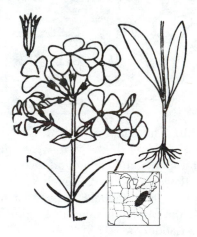

Figure 375

This unbranched plant rises to a height of 1 to 1 2/3 feet (30-51 cm). The leaves are 2 to 4 inches (5-10 cm) long, the upper ones sessile while the lower ones are obovate and petiolate. The broad corolla lobes are entire and pink or red. It is a woods plant and pretty much restricted to the mountains.

6a Basal leaves and those of the sterile shoots obovate to spatulate; sterile shoots creeping, rather numerous (Fig. 376) .... CREEPING PHLOX, *Phlox stolonifera* Sims

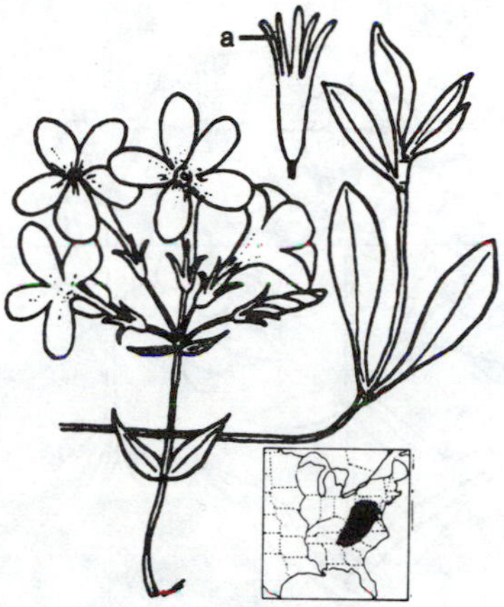

Figure 376

Mountain woods. This pubescent plant is much branched from its base into erect flowering shoots 4 to 10 inches (10-25 cm) high and creeping sterile shoots of considerable length. The longer lower leaves may attain a length of 3 inches (7.5 cm). The flowers are purple, violet or pink. The lobes of the corolla are rounded and seldom notched. The calyx lobes (a) are long and somewhat spreading.

    The western part of the country has several species of peculiar little phloxes adapted for living in the rigorous conditions of deserts or high altitudes. Three of these occur in dry soil in the area between the Rocky Mountains and the Missouri river. MOSS PHLOX, *P. bryoides* Nutt., is white-woolly and densely tufted. It has a height of only 2 or 3 inches. (2-7.5 cm) DOUGLAS' PHLOX, *P. douglasii* Hook., has subulate leaves up to an inch in length. HOOD'S PHLOX, *P. hoodii* Richards, is similar to *douglasii* but can be distinguished by its longer calyx which exceeds the corolla tube. These two species have a maximum height of 4 or 5 inches (10-13 cm) and are much less hair-covered than *bryoides*.

6b Basal leaves and those of the sterile shoots oblong or ovate (Fig. 377) .... WILD BLUE PHLOX, *Phlox divaricata* L.

Figure 377

One of the most common species of *Phlox* in its range, this plant grows in moist woods and thickets. It is 9 to 18 inches (23-46 cm) high, with flowers about an inch (2.5 cm) long. The corolla varies from a bluish-violet to almost white. The lobes are notched or entire and the flower 3/4 to 1¼ inch (1.9-3.2 cm) in diameter. The calyx lobes (a) are linear, as long as, and often longer than the tubular portion. The styles (b) are separate for about half of their length.

7a Lobes of the corolla divided (cleft) to the middle (Fig. 378) .. CLEFT PHLOX, *Phlox bifida* Beck

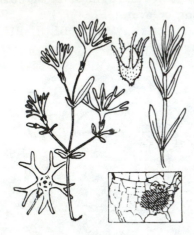

Figure 378

This much-branched, pubescent plant attains a height of 4 to 12 inches (10-30 cm). The linear leaves may be 2 inches (5 cm) long and the slender pedicels sometimes an inch (2.5 cm). The flowers are light purple; the cleft corolla-lobes make its identification simple. Like many narrow-leaved plants it grows in dry places.

CHICKWEED PHLOX, *P. bifida* ssp. *stellaria* (Gray) Wherry, growing on cliffs from Illinois to Tennessee, also has its lobes definitely cleft. The flowers are pale blue to almost white. ★

7b **Lobes of the corolla entire, not at all cleft** **8**

8a Plants almost wholly glabrous; calyx teeth subulate-lanceolate (a), shorter than the tube (Fig. 379) .... SMOOTH PHLOX, *Phlox glaberrima* L.

Figure 379

Prairies. The unbranched stems of this plant rise to a height of 4 feet (122 cm) or less. The upper leaves are lanceolate and may be 3 inches (7.5 cm) or more in length; the lower leaves are linear and usually shorter. The flowers are pink and have broadly-rounded corolla lobes which spread to 3/4 inch (2 cm) in diameter.

8b Plant downy or hairy, often glandular; calyx lobes long, linear, the tube short (Fig. 380) .... PRAIRIE PHLOX, *Phlox pilosa* L.

Figure 380

Dry soil. The entire plant is 1 to 2 feet (30-61 cm) high with slender stems and soft downy leaves. The leaves are linear or lanceolate and 1 to 3 inches (2.5-7.5 cm) long. The corolla is pink, purple or white with entire lobes. Another name is Downy Phlox.

HAIRY PHLOX, *P. amoena* Sims, with linear, often obtuse leaves and sessile, pink or white flowers, grows in dry places, Virginia to Kentucky, Alabama and Florida. Its leaves stand erect instead of spreading as in *pilosa*.

HYDROPHYLLACEAE

WATERLEAF FAMILY

1a Flowers solitary, opposite the leaves; stamens included (Fig. 381) 2

1b Flowers several to many in an inflorescence; stamens exserted (Fig. 384) 3

2a Calyx with reflexed appendages in each sinus between the lobes (Fig. 381a) .... SMALL-FLOWERED NEMOPHILA, *Nemophila aphylla* (L.) Brummit

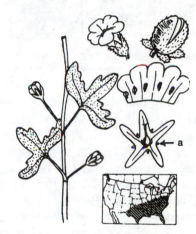

Figure 381

Synonym: *Nemophila microcalyx* (Nutt.) Fisch. & Mey.

Damp woods. The stems of this very slender, much-branched plant are from 2 to 16 inches (5-40 cm) long. The leaves are cut into 3 to 5 parts and may be 2 inches (5 cm) or more in length. The slender pedicels, 3/8 inch (1 cm) long, arise opposite the leaves. The tiny flowers may be either light blue or white. The corollas are only 1/8 inch (3 mm) long and the calyx shorter than the corolla.

2b Calyx without appendages; leaves pinnately divided into 7 to 13 narrow lobes (Fig. 382) .... NYCTELEA, *Ellisia nyctelea* L.

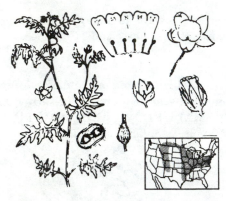

Figure 382

A very common rather inconspicuous plant of moist soil. It is a branched plant 2 to 16 inches (5-40 cm) high with pinnately divided leaves and small white or bluish flowers about 1/4 inch (6 mm) long. The calyx is nearly the same length as the corolla, but becomes enlarged to about 1 inch (2.5 cm) broad when in fruit. This flower is often found along the edge of gardens and other cultivated areas. The species is native to tropical America.

3a Leaves more than 3 inches (7.5 cm) wide; styles united (genus HYDROPHYLLUM) **4**

3b Leaves 3 inches (7.5 cm) wide or less; styles free for 1/2 to 1/4 their length (genus PHACELIA) **6**

4a Calyx with conspicuous reflexed appendages between the lobes (a); plants with tap roots (Fig. 383) .... APPENDAGED WATERLEAF, *Hydrophyllum appendiculatum* Michx.

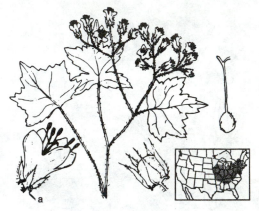

Figure 383

A pubescent plant of damp rich woods, this species is 1 to 2 feet (30-61 cm) high with the lower leaves pinnately divided, the upper ones ovate or orbicular, 5-lobed, and variously toothed. The violet or purple flowers about 1/2 inch (1.3 cm) long have slightly exserted stamens and are in a cyme. In the genus *Hydrophyllum*, there is a single appendage on the corolla between each stamen, but it is rolled inward and often appears as two (Fig. 385a).

4b Calyx without reflexed appendages; plants with fibrous roots **5**

5a Main stem leaves palmately lobed; inflorescence peduncles shorter than the petioles of the associated leaf (Fig. 384) .... CANADA WATERLEAF, *Hydrophyllum canadense* L.

5b Main stem leaves pinnately lobed; peduncles longer than the petioles of the associated leaf (Fig. 385) .... VIRGINIA WATERLEAF, *Hydrophyllum virginianum* L.

Figure 384

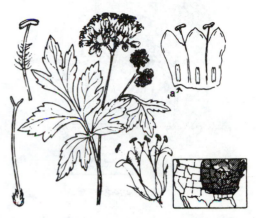

Figure 385

Moist woods. These plants with large shallowly or deeply 5- to 9-lobed maple-like leaves have stems 12 to 20 inches (30-50 cm) tall. The plants are glabrous or only slightly pubescent. The leaves grow to 10 inches (25 cm) broad. White to lavender campanulate flowers a half inch long (1.3 cm) or less are in loose cymes which are shorter than the leaves. The fruits of this and the other Waterleafs are globose capsules. Young leaves and shoot tips can be boiled 5 minutes and eaten as a vegetable.

Damp rich woods. This handsome plant is 1 to 3 feet (30-91 cm) high with pinnately divided, toothed leaves and white or violet flowers in cymes. At first the flower cluster is very dense, but gradually becomes more open. The corolla is 1/4 to 3/8 inch (6-10 mm) long with the 5 bearded stamens protruding beyond it. Rolled appendages (a) on the corolla alternate with the stamens. This blooms from May to August.

LARGE-LEAVED WATERLEAF, *H. macrophyllum* Nutt., is a somewhat coarser, larger plant than the preceding and in contrast with it is covered with a heavy coat of mixed soft and stiffer hairs and in the leaf segments ending rather bluntly. It ranges from Virginia to Illinois and southward to Tennessee and Alabama. The flowers are white or whitish, about 1/2 inch (1.3 cm) long and have long stamens.

6a Corolla lobes fringed (Fig. 386) ....
.... PURSH'S PHACELIA,
Phacelia purshii Buckley

6b Corolla lobes entire (Fig. 387) ....
...... LOOSE-FLOWERED PHACELIA,
Phacelia bipinnatifida Michx.

Figure 386

Figure 387

This pubescent annual ranges in height from 4 to 20 inches (10-50 cm). The leaves are pinnately cut, the lower ones having petioles while those higher are often sessile and clasping. The one-sided racemes bear 8 to 20 blue or white flowers about 1/2 inch (1.3 cm) across. The fringed corolla lobes make its identification easy.

FRINGED PHACELIA, *P. fimbriata* Michx., growing in the southeastern mountains, is similar but has obtuse calyx and leaf tips. This is not a common species.

Moist open woods and shady banks. This Phacelia grows 4 inches to 2 feet (10-61 cm) high with pinnately divided 3-inch (7.5 cm) leaves and blue-violet flowers, 3/8 to 1/2 inch (1-1.3 cm) broad, that are in loose racemes. The corolla bears 2 conspicuous appendages between each stamen (a).

SMALL-FLOWERED PHACELIA, *P. dubia* (L.) Trel., grows in moist soil from New York to Georgia and Alabama. It has leaves up to 2 3/8 inches (6 cm) long and small flowers 3/8 inch (1 cm) across with pubescent filaments.

There are 8 species of *Phacelia* listed for the eastern states. In western mountains, foothills, and valleys there are many additional species.

BORAGINACEAE

BORAGE FAMILY

1a **Lower leaves 4 to 12 inches (10-30 cm) long with long petioles** **2**

1b **Lower leaves 3 1/2 inches (9 cm) long or less, sessile or with short petioles** **4**

2a **Plant glabrous; corolla funnelform with a long tube and an entire limb; calyx much shorter than corolla; buds pink, turning blue upon opening (rarely white) (Fig. 388)** .... **BLUEBELL,** *Mertensia virginica* **(L.) Pers. ex Link**

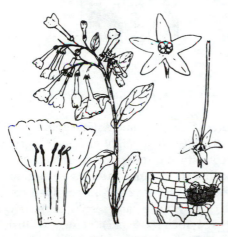

Figure 388

A showy plant of moist woods, especially on flood plains. It grows from 12 to 28 inches (30-70 cm) tall with a glabrous stem and leaves. The leaves are long-petiolate low on the stem and gradually become smaller and nearly sessile higher up. They are elliptic to oblanceolate or obovate, the lower ones to 8 inches (20 cm) long and the middle ones 6 inches (15 cm). The flowers are 3/4 to 1 inch (1.8-2.5 cm) long in terminal cymes. The corolla tube is longer than the limb; the stamens are inserted at the junction with the limb. Often this flower forms vast patches of blue and is then

a dazzling sight. The flowers alone do not smell fragrant but a field of them has a sweet somewhat chocolate-like smell. The flowers are visited by bumblebees and honeybees. There is also a mutant white-flowered form. Another name is Virginia Cowslip.

Wet meadows and mountains in the western states have their own species of Bluebells.

2b **Plant pubescent; corolla limb with 5 separate lobes; calyx at least 1/3 as long as the tube** **3**

3a **Flowers blue, occasionally white; inflorescence terminal on a leafless stalk (Fig. 389)** .... **WILD COMFREY,** *Cynoglossum virginianum* **L.**

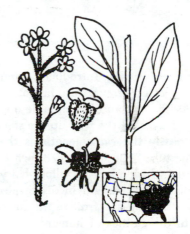

Figure 389

Dry woods. This perennial has a sturdy, unbranching stem 1 to 2½ feet (30-76 cm) or more in height, and is hirsute. The oval, entire, lower leaves are from 4 inches to a foot (30 cm) in length. The upper leaves are few, smaller, sessile and with cordate, clasping bases. The flowers, 1/2 inch (1.2 cm) across, are blue or occasionally white. The fruit is 4 nutlets (a) each about 1/4 inch (6 mm) wide and covered with stiff hooked bristles.

3b Flowers dull red, occasionally white; inflorescences in the axils of leaves to the top of the stem (Fig. 390) .... HOUND'S-TONGUE, *Cynoglossum officinale* L.

5a Flowers solitary in the axils of the upper leaves; corolla tube seldom longer than the calyx, often shorter (Fig. 391) .... CORN GROMWELL, *Buglossoides arvense* (L.) I. M. Johnston

Figure 390

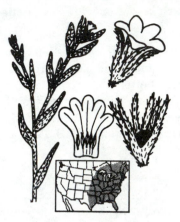

Figure 391

Synonym: *Lithospermum arvense* L.

Roadsides and fields. Introduced from Europe, this species grows to a height of 1 to 4 feet (30-120 cm) and has sizeable oblong to lanceolate leaves all the way up the stem. The calyx is nearly the same length as the dark red corolla tube. The limb is 3/8 inch (1 cm) broad. The whole plant is villous-hairy. The bristles on the nutlets aid in dispersal of the seeds by attaching the fruit to the fur of animals or to the clothes of humans.

This erect European and Asiatic plant, growing in waste places to a height of 2½ feet, is thickly covered with an appressed pubescence. Most of the linear to lanceolate leaves are sessile and measure 3/4 to 2 3/8 inches (2-6 cm) in length; the flowers are sessile or nearly so, about 1/4 inch (0.6 cm) across, whitish or pinkish. The glabrous brown nutlets are wrinkled and pitted.

4a Flowers blue or white **5**

4b Flowers yellow **7**

5b Flowers in terminal, coiled, racemose inflorescences **6**

6a Corolla blue with a yellow eye; plant of brook banks and marshes (Fig. 392) ....
.... **FORGET-ME-NOT,**
Myosotis scorpioides **L.**

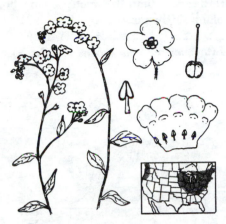

Figure 392

Brook banks and wet soil. A native of Europe that has escaped from gardens and become naturalized. The plant is erect or decumbent, 8 to 28 inches (20-70 cm) long, with pubescent foliage and loose, many-flowered inflorescences which are often curved like a question mark. The corolla, 1/4 to 3/8 inch (6-10 mm) broad, is blue with a yellow eye.

Our native SMALLER FORGET-ME-NOT, *M. laxa* Lehm., ranging from Ontario to Virginia and Tennessee and North Carolina contrasts with the above by having a corolla no longer than the calyx and a corolla limb less than 1/4 inch (6 mm) wide and is thus less conspicuous.

6b Corolla white; plant of dry areas (Fig. 393) .... **SPRING SCORPION-GRASS,** *Myosotis verna* **Nutt.**

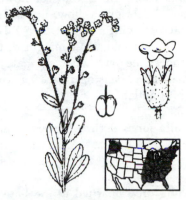

Figure 393

Dry banks and rocky woods. The plant grows 2 to 16 inches (5-40 cm) tall with hispid foliage and curved racemes of white flowers less than 1/8 inch (3 mm) broad. Three of the calyx lobes are shorter than the other two.

7a Corolla lobes wavy-margined and toothed; flowers about an inch (3 cm) long, bright or pale yellow (Fig. 394) ....
.... **NARROW-LEAVED PUCCOON,**
Lithospermum incisum **Lehm.**

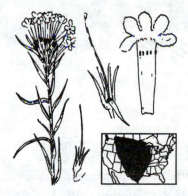

Figure 394

Dry or sandy soil. An attractive plant, often branching near the base, growing on the prairies of Ontario and Indiana west and south to British Columbia, Texas, Utah and Arizona. It grows 2 to 20 inches (5-40 cm) high from a woody taproot. The leaves are linear and pubescent. There are 2 kinds of flowers; the early ones are bright yellow, up to 3/4 inch (2 cm) across, with a tube 1/2 to 1 1/8 inch (1.5-3 cm) long and wavy-margined, toothed corolla lobes; the later ones are pale yellow, much smaller and cleistogamous. The fruits are shiny white nutlets.

7b Corolla lobes entire; flowers orange yellow **8**

8a Corolla tube bearded at the base within (a); flower 1/2 to 1 inch (1.5-2.5 cm) broad; plant roughly hairy (Fig. 395) .... **HAIRY PUCCOON,** *Lithospermum carolinense* (**J. F. Gmel.**) **MacM.**

Figure 395

Synonym: *L. croceum* Fern.

This stout plant is found in dry woods and sandy soil. It grows 1 to 2½ feet (30-76 cm) high with rough-pubescent stems, linear to lanceolate leaves, and bright orange-yellow flowers 1/2 to 2/3 inch (1.3-1.7 cm) long. The ovary consists of 4 deeply-parted lobes or nutlets, but often only one nutlet matures.

The plant blooms from April to June. *Lithospermum* comes from the Greek meaning stone and seed, in reference to the hard seed.

8b Corolla tube not bearded at the base; flower less than 3/8 inch (1 cm) broad; plant softly pubescent (Fig. 396) .... **HOARY PUCCOON,** *Lithospermum canescens* (**Michx.**) **Lehm.**

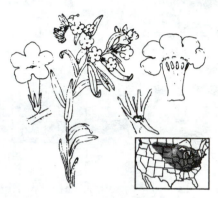

Figure 396

The Hoary Puccoon is found in open woods, on plains and sandy soil of southeastern Canada and in much of the United States. It is rare in New England. The plant grows 4 to 20 inches (10-51 cm) high, with hirsute-pubescent foliage and bright orange-yellow flowers 1/2 inch (1.3 cm) broad in dense leafy inflorescences. Its nutlets are smooth, shining and white.

VERBENACEAE

VERVAIN FAMILY

1a Flowers large, more than 3/8 inch (10 mm) across; bracts of spike shorter than or equalling the calyx (Fig. 397) .... ROSE VERBENA, *Verbena canadensis* (L.) Nutt.

1b Flowers small, 1/8 inch (3 mm) across; bracts much longer than the calyx (Fig. 398) .... PROSTRATE VERVAIN, *Verbena bracteata* L. & R.

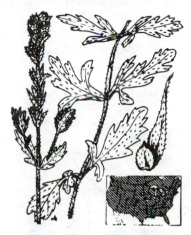

Figure 398

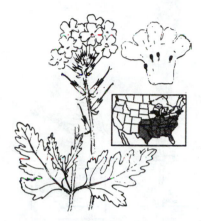

Figure 397

Dry soil of open woods and prairies. The plant has pubescent stems, erect or decumbent, from 4 to 24 inches (10-60 cm) long. The basic leaf shape is ovate to lanceolate, the margins are variously lobed and rounded- or pointed-toothed. The purplish red, white or blue flowers are in dense capitate spikes. The floral tube is nearly 1 inch (2.5 cm) long and the limb 1/2 inch (1.3 cm) broad. The glandular calyx is up to 1/2 inch (1.3 cm) long with its associated bract as long or shorter.

The cultivated GARDEN VERBENA, *V.* X *hybrida* Voss, is similar to this species and occasionally escapes from gardens. The red, violet, blue (and white) flowers often have a white eye. The leaves are lanceolate, densely pubescent, with crenate or serrate margins.

This branching, prostrate to ascending, hirsute plant is very common on prairies and in barnyards and other waste places. The stems with deeply lobed leaves may attain lengths of 20 inches (50 cm) and are terminated by a flowering spike which keeps elongating until it is sometimes 8 inches (20 cm) long. The tiny flowers are purplish blue, and shorter than the leaflike or linear bracts.

About 200 species of *Verbena* are known; most of them are native in the tropics and subtropics of North and South America.

LABIATAE [LAMIACEAE]

MINT FAMILY

1a Calyx with a protuberance on the upper side (a); calyx 2-lipped, the lips entire (Fig. 399) .... SHOWY SKULLCAP, *Scutellaria serrata* Andr.

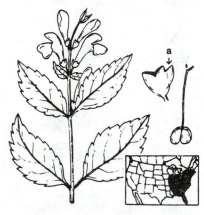

Figure 399

A handsome woodland plant growing 1 to 2 feet (30-60 cm) high. The inch-long flowers (2.5 cm) are blue and are in a loose terminal inflorescence. The leaves are ovate to elliptic, 1½ to 5 inches (4-13 cm) long and petiolate. The calyx has a protuberance on the upper side.

SMALL SKULLCAP, *Scutellaria parvula* Michx., grows 4 to 12 inches (10-30 cm) high in sandy soil in southeastern Canada and in most of the eastern half of the United States. The purplish flowers are 1/4 to 3/8 inch long (0.6-1 cm) and are solitary in the leaf axils. Its leaves at mid-stem are nearly sessile, ovate, about 1/2 inch (1.3 cm) long.

1b Calyx without a protuberance on the upper side; 2-lipped or nearly regular, 3- to 5-toothed 2

2a Anther-bearing stamens 2 3

2b Anther-bearing stamens 4 4

3a Flowers in the axils of the leaves; leaves linear, to 3/4 inch (2 cm) long (Fig. 400) .... MOCK PENNYROYAL, *Hedeoma hispidum* Pursh

Figure 400

Sandy places, dry soil. The pubescent stems have leaves and flowers along most of their 2 to 8 inch (5-20 cm) length. The leaves are sessile, linear, to 3/4 inch (2 cm) long, and usually have a cluster of a few flowers in the axil. The blue flowers have a pubescent, bilabiate calyx, both lips with long teeth. The two stamens curve upward along the upper corolla lip. The fruits are smooth nutlets.

3b Flowers in a leafless terminal inflorescence; leaves mostly basal, some of them lyrate pinnatifid (Fig. 401) .... **LYRE-LEAVED SAGE,** *Salvia lyrata* L.

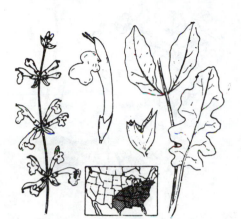

Figure 401

Dry sandy woods and barrens. The plant is 1 to 2½ feet (30-76 cm) high with lyre-shaped basal leaves and conspicuous lavender flowers about 1 inch (2.5 cm) long that are arranged in distant whorls. The upper lip of the corolla is much smaller than the lower one which is broadly 3-lobed and forms a convenient landing place for insect visitors. The calyx is bilabiate with 2 deep lobes on the bottom and 3 shallower lobes on the top.

This is an immense genus with some 500 species in temperate and tropical regions. Sage, the well known seasoning is *S. officinalis* L., and one of the most brilliant summer plants is SCARLET SAGE, *S. splendens* Ker-Gawl., in which both the corolla and calyx are bright scarlet.

4a Calyx 2-lipped, 3- to 5-toothed; leaves ovate or ovate-lanceolate; flowers in dense terminal spikes (Fig. 402) .... **HEAL-ALL,** *Prunella vulgaris* L.

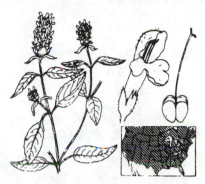

Figure 402

An introduced plant that is now common in fields, woods, and waste places in most of North America. It grows 4 to 32 inches (10-80 cm) high with ovate to lanceolate petiolate leaves and dense spikes of purplish or white flowers 3/8 to 3/4 inch (1-2 cm) long. The flower spikes are 1/2 to 1 inch (1.3-2.5 cm) long, but later become 2 to 4 inches (5-10 cm) in length. The dried fruit heads of this plant are conspicuous all winter.

Another version of the name is Self-heal. Both of the names refer to early medicinal use of the plant as an "aromatic" or "bitter" and an astringent. Many members of the mint family contain oils such as menthol which give a cool feeling to the mouth and a cool tingling sensation to the skin.

4b Calyx nearly regular, 5-toothed; leaves palmately lobed or orbicular to cordate; flowers in axillary or few-flowered terminal clusters 5

5a Leaves palmately divided, the lobes with coarse teeth; flowers in axillary clusters well separated from other clusters (Fig. 403) .... MOTHERWORT, *Leonurus cardiaca* L.

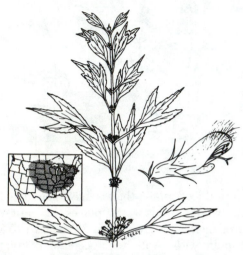

Figure 403

Naturalized in fields and roadsides. This coarse perennial may have a number of stems branching from the base, 1 1/3 to 5 feet (40-150 cm) high, and bearing leaves 2 to 4 3/4 inches (5-12 cm) long. The leaves become progressively smaller toward the top and bear clusters of flowers in the axils. The calyx is nearly regular, the 5 lobes stiff and spiny, about as long as the tube. The corolla is purplish to white, about 3/8 inch (1 cm) long. The nutlets are flattened and pubescent on the top. The scientific name refers to its use in ancient times as a heart treatment. The common name comes because it is used as a medicine at childbirth.

5b Leaves not lobed, round or heart shaped in outline, the margins crenate **6**

6a Plants creeping; upper pair of stamens longer than the lower ones; flowers blue or violet (Fig. 404) ...... GROUND IVY, *Glechoma hederacea* L.

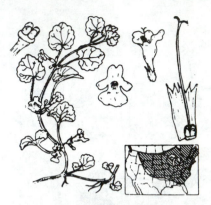

Figure 404

Damp woods, waste places, and around dwellings. This mint has creeping stems which root at the nodes and erect flowering stems. The leaves are round or reniform with crenate margins and long petioles. The violet or blue flowers, 1/2 to 1 inch (1.3-2.5 cm) long, are in few-flowered clusters in the leaf axils. The upper lip is smaller than the lower, which has a broad middle lobe and narrower lateral ones. The plant is native of Eurasia but is now widespread in moist soil. Leaves have been used in tea, tonics, and salves. There are many picturesque common names such as Gill-over-the-ground and Robin-running-in-the-hedge. Bees and moths search the flowers for nectar.

6b Plants decumbent; upper pair of stamens shorter than the lower ones; flowers pink to purple **7**

7a Leaves subtending the flower clusters sessile and clasping (Fig. 405) .... **HENBIT,** *Lamium amplexicaule* L.

7b Leaves subtending the flower clusters petiolate (Fig. 406) .... **DEAD NETTLE,** *Lamium purpureum* L.

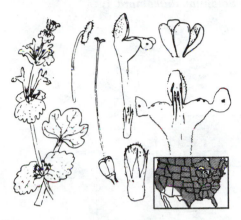

Figure 405

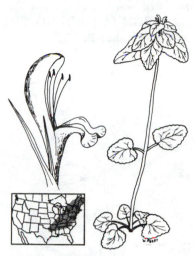

Figure 406

This species is a common plant 4 to 18 inches (10-46 cm) of waste areas and cultivated ground that was introduced from Europe and is now established in southeastern Canada and in most of the United States. The coarsely-crenate leaves are orbicular with the upper ones nearly sessile and the lower ones on petioles. The purplish-red flowers are 1/2 to 2/3 inch (1.3-1.7 cm) long. The upper lip is quite reddish and is pubescent; the lower lip is spreading and is spotted with dark purple. Sometimes the flowers are cleistogamous.

Naturalized from Eurasia in fields, gardens, and roadsides. The 3 inch to 1 foot (8-30 cm) stems have petiolate leaves and end with a leafy crowded spike of purple or pinkish flowers. The leaves are rounded or heart shaped, pubescent and petiolate, and often become larger higher on the stem. The leafy bracts around the flowers are similar to the largest leaves and often tinged with purple. The flowers are about 5/8 inch long (1.5 cm) with a calyx about a third as long. The nutlets have oil bodies attached to them. These oil bodies are carried away by ants and the seeds are dispersed in the process.

SOLANACEAE

NIGHTSHADE FAMILY

1a Corolla funnelform, shallowly lobed, yellowish (Fig. 407) .... VIRGINIA GROUND CHERRY, *Physalis virginiana* P. Mill.

Figure 407

1b Corolla spreading to reflexed, deeply lobed, purple (Fig. 408) .... BITTER NIGHTSHADE, *Solanum dulcamara* L.

Figure 408

In sun in woods and fields. This is an erect plant, perennial from a rhizome, with stems 1 to 2 feet (30-60 cm) tall. The leaves are lanceolate to ovate with entire or sinuately toothed margins. The flowers are nodding. The corolla is funnel shaped, 1/2 to 3/4 inch (1.3-2 cm) long, light yellow with 5 dark purplish spots at the base inside. In flower the calyx is only 3/8 inch (1 cm) long but as the fruit matures the calyx tube expands to 1¼ inch (3.3 cm) long and 3/4 inch (2 cm) broad forming a balloon around the half inch (1.3 cm) berry. This inflated calyx is characteristic of the whole genus.

The CLAMMY GROUND CHERRY, *P. heterophylla* Nees, blooms in May, and grows from Quebec to Minnesota to Texas and Georgia. The stems and leaves are pubescent with glandular hairs which give a clammy feel to the plant.

Naturalized from Eurasia. This *Solanum* is a vine capable of climbing or trailing 3 to 9 3/4 feet (1-3 m). The leaves are ovate, 1 to 4 inches (2.5-10 cm) long, sometimes with one or two small lobes at the base. The flowers which are borne on an open panicle are only 1/2 inch (1.3 cm) long but quite showy. The corolla is usually purple and reflexed, making a striking contrast with the 5 bright yellow elongate anthers grouped together in the center of the flower. The bright red berries have caused this vine to be called Bittersweet, but the true Bittersweet is a woody vine in a completely different family.

HORSE NETTLE, S. *carolinense* L., blooms in late spring in the Southeast. It is a coarse plant with viciously spiny stems and leaf veins. The flowers are pale violet or white with a spreading corolla an inch (2.5 cm) or more broad. The fruit is a yellow berry. ★—FL ★—GA

SCROPHULARIACEAE

FIGWORT FAMILY

1a Leaves on stem alternate 2

1b Leaves opposite 4

2a Corolla spurred on the lower side near the base (a) (Fig. 409) ....
.... **BLUE TOADFLAX,**
Linaria canadensis (L.) Dum.-Cours.

Figure 409

This very slender plant of dry sandy soil grows 4 to 30 inches (10-76 cm) high and has shiny linear leaves and small bluish flowers 1/4 to 1/3 inch (6-8 mm) long, in long racemes. The flowers are bilabiate, the lower lip with 3 flaring lobes.

BUTTER-AND-EGGS, *L. vulgaris* P. Mill. is a more conspicuous relative that has orange and yellow flowers that are at least an inch (2.5 cm) long. This species blooms in May but is at its best in the summer and fall. It closely resembles the cultivated Snapdragon except for the sharp spur.

2b Corolla without a spur 3

3a Leaves usually deeply 3- to 5-cleft from the tip, occasionally entire; bracts bright red, conspicuous (Fig. 410) ....
.... **INDIAN PAINTBRUSH,**
Castilleja coccinea (L.) Spreng.

Figure 410

Fields and moist woods. The plant is 4 to 28 inches (10-70 cm) tall with entire basal leaves and alternate, usually divided, leaves on the stem. The leaf is divided from the tip into 3 to 5 linear lobes up to 1½ inch (4 cm) long. The flowers are greenish yellow and about an inch (2.5 cm) long in a dense spike. The vivid scarlet of the tips of the 3- to 5-lobed bracts (a) provides the color. The bracts are equal to or longer than the flowers and make the inflorescence look somewhat like an artist's paintbrush dipped in scarlet paint. Another name is Scarlet Painted-Cup.

Castilleja sessiliflora Pursh, DOWNY PAINTED-CUP, blooms on the plains of the midwest in April to June as well. Its flowers are lemon yellow or whitish and curved outward over the shorter green bracts. The plant is softly hairy.

Numerous species of this genus grow in the western mountains of North America.

3b Leaves pinnately parted and toothed; bracts green, leaf-like (Fig. 411) .... WOOD BETONY, *Pedicularis canadensis* L.

Figure 411

Dry woods. This common plant grows 4 to 16 inches (10-40 cm) high with mostly basal leaves that are pinnately parted into oblong, obtuse, dentate lobes. The yellow or reddish flowers, 1/2 to 5/6 inch (1.3-2 cm) long, are in a crowded capitate spike that becomes elongated in fruit. It is also known as Common Lousewort and Beefsteak Plant.

There are only 4 species of this genus in the eastern United States, at least one of them very rare, but 14 species have been identified in the Pacific Northwest.

4a Anther bearing stamens 2 5

4b Anther bearing stamens 4; corolla 2-lipped, 4- to 5-lobed 11

5a Corolla rotate, usually 4-lobed, tube short; sepals 4 (genus VERONICA) 6

5b Corolla bilabiate, tubular; sepals 5 (Fig. 412) .... HEDGE HYSSOP, *Gratiola neglecta* Torr.

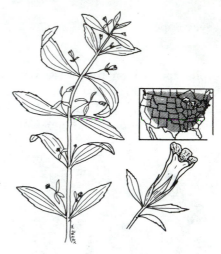

Figure 412

Moist soil. Pubescent stems, 4 to 16 inches (10-40 cm) tall, carry lanceolate to oblanceolate, opposite leaves and bracts. The flowers are solitary in the axils of leaf-like bracts on long slender pedicels. The bilabiate corolla, 3/8 inch (1 cm) long, has a yellowish tube and white lobes.

6a Flowers borne in racemes which arise from the axils of paired leaves 7

6b Flowers solitary in the axils of alternate, bract-like leaves 8

7a Plants pubescent, growing in dry soil (Fig. 413) .... COMMON SPEEDWELL, *Veronica officinalis* L.

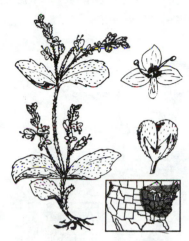

Figure 413

Dry fields and woods. This pubescent, decumbent perennial growing from 2 to 10 inches (5-25 cm) high has short petioled elliptic to ovate leaves 3/4 to 2 inches (1.5-5 cm) long with coarsely toothed margins. The flowers are pale blue, about 1/4 inch (5 mm) broad on crowded racemes. The fruit is flattened, obcordate, and pubescent. Tea from the plants is a tonic or a cough medicine.

7b Plants glabrous, growing in swamps and shallow water or along streams (Fig. 414) .... WATER SPEEDWELL, *Veronica anagallis-aquatica* L.

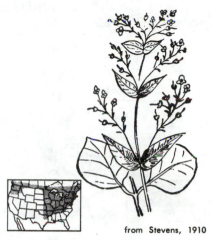

from Stevens, 1910

Figure 414

This is an erect or decumbent plant rooting at the nodes in wet soil. When erect, the stems can reach to 2 or 3 feet (60-100 cm). The glabrous leaves are lanceolate to elliptic and up to 4 inches (10 cm) long. They are all opposite, and either sessile or clasping. The racemes are usually borne one in each axil of a pair of leaves. Each long raceme consists of many pale violet-blue flowers 1/4 inch (5 cm) broad. The fruit is only minutely notched at the apex. This species is native in Eurasia but has now spread nearly worldwide to ditch and stream banks from South Africa, Egypt, and Tibet to all across North America. It has been recommended in Europe as a source of Vitamin C.

The native AMERICAN BROOKLIME, *Veronica americana* (Raf.) Schwein., is similar and grows in the same habitats. Its leaves are lanceolate to ovate and with short petioles.

8a Flowers not in distinct terminal racemes or spikes, appearing axillary; the bracts leaflike, hardly different from the opposite stem leaves; pedicels longer than the associated bracts (Fig. 415) .... BIRD'S-EYE, *Veronica persica* **Poir.**

Figure 415

Roadsides, lawns and gardens. The stems of this pubescent annual spread into a rosette from the base and attain a length of 4 to 16 inches (10-40 cm). The leaves are ovate, short petioled and up to 1 inch (2.5 cm) long with crenate or serrate margins. The flowers, about 3/8 inch (1 cm) across are blue, darker on the upper lobe and paler on the lower, with dark blue veins. There is a white eye in the center. The fruit (a) is wider than high, the two lobes spreading.

8b Flowers in distinct terminal racemose inflorescences; pedicels shorter than the bracts **9**

9a Bracts suddenly reduced from the foliage leaves; pedicels about 1/8 inch (2-5 mm) long (Fig. 416) .... **THYME-LEAVED SPEEDWELL,** *Veronica serpyllifolia* **L.**

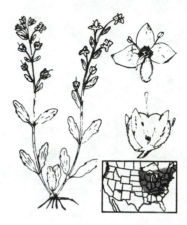

Figure 416

The downy-pubescent stems of this perennial arise to a height of 2 to 12 inches (5-30 cm). The ovate leaves are seldom over 1/2 inch (1.3 cm) long; the flowers are from 1/6 to 1/3 inch (4-8 mm) across and are white or blue with darker stripes. It seems to be native of our country although it is also found in Europe.

9b Bracts gradually reduced from the foliage leaves; flowers sesssile or nearly so **10**

10a Corolla white; leaves oblanceolate to linear 1/4 to 1 1/2 inch long (0.5-3 cm), glabrous (Fig. 417) .... **PURSLANE SPEEDWELL,** *Veronica peregrina* L.

10b Corolla blue; leaves ovate to elliptic, 1/2 inch (1.2 cm) long, pubescent (Fig. 418) .... **CORN SPEEDWELL,** *Veronica arvensis* L.

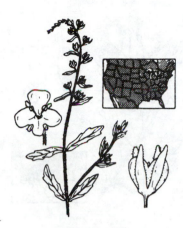

Figure 417

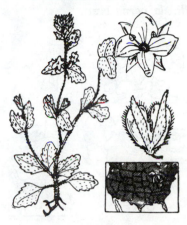

Figure 418

A common plant of moist soil found in waste areas and cultivated grounds. The plant is 2 to 12 inches (5-30 cm) high with tiny white flowers 1/12 inch (2 mm) broad. The capsule is emarginate. The plants grow erect. It is widely distributed not only in North America but also in South America and Eurasia. Another common name is Neckweed.

This European native grows to a height of 2 to 12 inches (5-30 cm). The lowest leaves are broadly oval with fairly long petioles while the upper leaves are nearly sessile. The flower pedicels are shorter than the calyx. The flowers are blue or bluish. The species grows in woods, waste places and fields and is known from several continents. Small sweat bees visit these flowers.

11a (4) Flowers in a raceme or panicle 12

11b Flowers axillary, the middle lobe of the lower lip folded together and enclosing the stamens and the style (a); upper lip whitish, lower lip blue (Fig. 419) .... BLUE-EYED MARY, *Collinsia verna* Nutt.

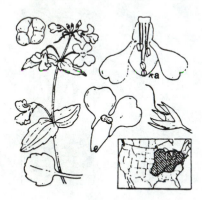

Figure 419

A beautiful and delicate flower of moist woods. The plant grows 8 to 20 inches (20-51 cm) high with blue and white flowers, 1/2 to 2/3 inch (1.3-1.7 cm) long, in the leaf axils. The main leaves are toothed, ovate, and broadest at the base; the basal ones with long petioles.

NARROW-LEAVED COLLINSIA, *Collinsia violacea* Nutt., found from Illinois to Kansas and south to Texas and Arkansas, is similar but has mostly entire leaves narrowed at the base and the lower lip of the flower violet.

12a Flowers greenish or brownish, inflorescence much-branched, open; one stamen sterile, reduced, not bearded (Fig. 420) .... HARE FIGWORT, *Scrophularia lanceolata* Pursh

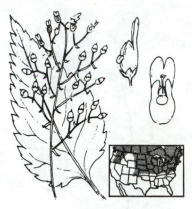

Figure 420

Open woods and roadsides. This plant is 3 to 6½ feet tall (1-2 m) with ovate or lanceolate leaves with coarse, sharp teeth. The 4 to 12 inch (10-30 cm) panicles have many small flowers. The corolla has 2 reddish-brown upper lobes as long as the tube, 2 shorter lateral lobes and a lower lobe that is yellowish green and reflexed. There are 4 fertile stamens and a reduced yellow-green sterile stamen.

MARYLAND FIGWORT, S. *marilandica* L., with range similar to the above but blooming somewhat later can be distinguished by its dull (not shining) corolla and the sterile stamen being purple instead of greenish yellow.

12b Flowers white to purple; the one sterile stamen equalling the 4 fertile ones and bearded toward the tip (genus PENSTEMON) 13

13a Flowers lavender or white with purple stripes, 1 3/8 to 2 3/8 inch (3.5-6 cm) long 14

13b Flowers white, 1 1/8 inch (3 cm) long (Fig. 421) .... **FUNNELFORM BEARD-TONGUE,** *Penstemon tubiflorus* Nutt.

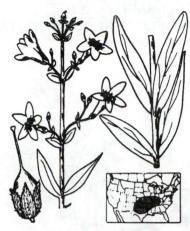

Figure 421

14a Leaf margins entire; leaves thick, glabrous and glaucous (Fig. 422) .... **LARGE-FLOWERED BEARD-TONGUE,** *Penstemon bradburii* Pursh

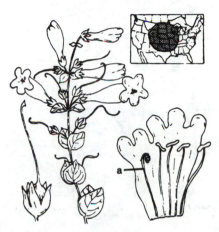

Figure 422

This moist-soil plant grows to a height of 2 to 3 feet or more (30-100 cm) and has a sticky pubescence on its upper parts but is largely glabrous below. The pedicels are short. The white corollas, almost 1 inch (2.5 cm) long, have a wide, funnel-like throat. The capsule, not counting the persistent style, is about twice as long as the sepals. The leaves are oblanceolate to oblong.

There are 230 species of *Penstemon* listed as growing in North America. Many of them are in the West. Beard-tongues are said to get their common name from the single hairy stamen.

Synonym: *Penstemon grandiflorus* Nutt.

A beautiful prairie plant which has been introduced eastward from its more western range. The lavender-blue corolla is 1½ to 2 inches (3.8-5 cm) long. The sterile filament is curved and slightly bearded. The leaves are entire; the basal ones obovate, the upper stem leaves sessile and oval. The entire plant, 2 to 4 feet (61-122 cm) high, is glabrous and somewhat glaucous and very attractive either growing wild or in cultivation.

14b Leaf margins dentate; the inflorescence pubescent (Fig. 423) .... **COBAEA BEARD-TONGUE,** *Penstemon cobaea* Nutt.

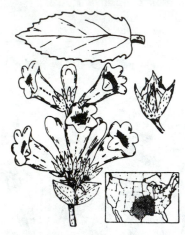

Figure 423

Dry prairies. The showy white to pale purple flowers with darker purple lines are 1¼ to 2¼ inches (3.5-6 cm) long and glandular hairy. The stems are 2¼ feet (70 cm) tall, pubescent in the inflorescence. The leaves are toothed, the lower ones spatulate or oblanceolate on petioles, the upper ones sessile and lanceolate to ovate.

HAIRY BEARD-TONGUE, *Penstemon hirsutus* (L.) Willd., also has toothed leaves, a glandular-hairy inflorescence, and glandular, pale violet corollas. The flowers are an inch (2.5 cm) long and the 1 to 3 foot (30-90 cm) stems are covered with fine white hairs. It blooms in dry woods and fields from Quebec to Wisconsin and into Virginia and Tennessee.

OROBANCHACEAE

BROOM-RAPE FAMILY

1a Flowers white or lavender, borne singly on naked peduncles; corolla regular; stamens included (Fig. 424) .... **ONE-FLOWERED CANCER-ROOT,** *Orobanche uniflora* L.

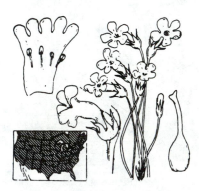

Figure 424

Cancer-root is a parasitic plant of damp woods and thickets. The short stem bears several brownish scales and 1 to 4 naked 1-flowered peduncles 2 to 8 inches (5-20 cm) tall. The flower, 1/2 to 3/4 inch (1.3-1.9 cm) long, is purplish or rarely white and is delicately fragrant. Another name is Naked Broom-rape. The plant has been applied medicinally as an astringent to ulcers and cancerous growths.

1b Flowers yellowish, in a dense spike; corolla irregular; stamens exserted (Fig. 425) .... SQUAW-ROOT, *Conopholis americana* (L.) Wallr.

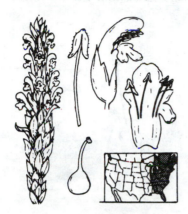

Figure 425

In dry woods in clusters at the base of trees (mostly oaks). This tan colored parasitic plant is 2 to 10 inches (5-25 cm) high and is covered with stiff scales so that it resembles a pine cone. The numerous pale yellow flowers, 1/2 inch long, are crowded in a dense spike. The upper lip of the corolla is entire; the lower lip is 3-lobed. Another name is Cancer-root.

PLANTAGINACEAE

PLANTAIN FAMILY

1a Leaves linear 2

1b Leaves ovate, cordate or lanceolate 3

2a Bracts much longer than the flowers; spike pubescent, not woolly (Fig. 426) .... LARGE-BRACTED PLANTAIN, *Plantago aristata* Michx.

Figure 426

Dry or sandy soil. Native in the United States west of Illinois but now growing over the eastern U.S., southeastern Canada and northern Mexico. The basal leaves are linear, not over 8 inches (20 cm) long and 3/8 inches (1 cm) wide, usually shorter than the scapes. The scapes are 6 to 10 inches (15-25 cm) tall, bearing flowers crowded on spikes 1 to 6 inches (2.5-15 cm) long. Linear bracts much exceed the flowers and fruits.

2b Bracts not much longer than the flowers; spike densely hairy (Fig. 427) .... **WOOLLY INDIAN WHEAT,** *Plantago patagonica* Jacq.

3a Spikes crowded at the ends of long scapes; leaves narrowly elliptic to lanceolate (Fig. 428) .... **ENGLISH PLANTAIN,** *Plantago lanceolata* L.

Figure 427

Figure 428

Synonym: *Plantago purshii* R. & S.

Native of dry plains and prairies in the western states and introduced into the East. The same species occurs in South America, as indicated by the scientific name. The leaves are shorter than the scape, 1½ to 6 inches (4-15 cm) long and only 1/4 inch (0.7 mm) wide. The inflorescences are 2 to 15 inches tall (5-38 cm) with woolly spikes 1 to 5 inches (2.5-12.5 cm) long composed of crowded flowers with persistent tannish corollas.

Psyllium seed is a bulk-producing laxative gathered from a Mediterranean *Plantago.* Seeds of *P. patagonica* are sometimes used as a substitute for psyllium.

This plant grows 4 to 24 inches (10-60 cm) high when flowering in fields, lawns, and waste places. The leaves are 4 to 16 inches long (10-40 cm) and 1/4 to 2 inches (0.7-5 cm) wide, either pubescent or glabrous. The flowers are in spikes 1/2 to 3 inches (1.5-8 cm) long on the ends of pubescent scapes. The spikes are at first oval and later become cylindrical. There is a green midrib in the otherwise translucent bracts and sepals. The stamens and stigmas extend far beyond the 1/8 inch (3 mm) corolla tube and its somewhat shorter spreading lobes.

Leaves can be eaten young as a pot herb and they have been used to stop bleeding from wounds. The mucilaginous seeds are a laxative. Mostly, however, this plant is recognized as a noxious perennial weed. When growing with red clover that is being cut for seed the light brown plantago seeds are so nearly the same size that separation is difficult.

3b Spikes elongate along the scape axis; leaves ovate to cordate with long petioles **4**

4a Plant of marshes and shallow streams; bracts and sepals with a narrow keel (Fig. 429) .... **HEART-LEAVED PLANTAIN,** *Plantago cordata* Lam.

4b Plant of dry, weedy sites; bracts and sepals prominently keeled (Fig. 430) .... **COMMON PLANTAIN,** *Plantago major* L.

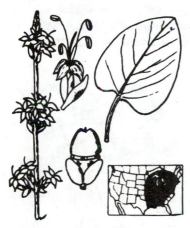

Figure 429

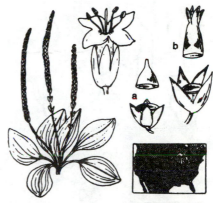

Figure 430

This species has another common name, Water Plantain, which is also a good one because it grows in and near streams and in swamps. The leaves are frequently as much as 10 inches (25 cm) long, but the flowering stems stand well above them, with the clusters of flowers scattered. The flowers are perfect; the capsules contain 1 to 4 flat-faced seeds and open near the middle. ★

This European weed of many common names is naturalized in lawns, along roadsides, and in weedy sites. The leaves are elliptic or ovate with long petioles. The spikes are 1½ to 8 inches (4-20 cm) long and densely flowered. The bracts are ovate. The capsule has many seeds and opens by a cap that separates from the seed cup (a) at about the middle. As with other plantains, this species too has had its leaves used for dressing wounds and bee stings, and as a pot herb, and its seed as a laxative.

RUGEL'S PLANTAIN, *Plantago rugelii* Dcne., is a native species in the eastern and midwestern United States that is very similar. In this species the petioles are purple at the base, and the elongate capsule (b) containing 4 to 10 seeds separates below the middle, about 1/3 of the way from the base.

RUBIACEAE

MADDER FAMILY

1a **Leaves opposite** **2**

1b **Leaves in whorls** **6**

2a Flowers in terminal pairs on a common peduncle; stigmas 4; leave ovate-orbicular (Fig. 431) .... **PARTRIDGE-BERRY,** *Mitchella repens* L.

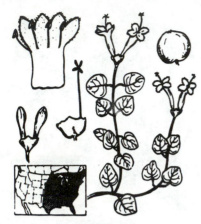

Figure 431

A trailing plant with broad shiny heart-shaped leaves to 3/4 inch (2 cm) long and dainty 4-lobed flowers borne in pairs. The stems are 4 to 12 inches (10-30 cm) long. The flowers, about 1/2 inch (1.3 cm) long, are white and finely-pubescent inside, smooth and tinged with pink or purple outside. The two flowers have their calyces united. The fruit is an edible, usually scarlet berry that remains all winter. Tea from the fruits was given to assist labor in childbirth, and tea from the leaves was used by some American Indians as a cure for insomnia. This plant is found in dry or moist woods and stream banks.

2b Flowers solitary or in cymes; stimas 2; leaves ovate to linear (genus **HEDYOTIS**) 3

3a Flowers solitary; corolla with lobes spreading perpendicular to the tube (salverform) (Fig. 432a); leaves shorter than 3/8 inch (1 cm) 4

3b Flowers in cymes; corolla funnelform; leaves longer than 3/8 inch (1 cm) 5

4a Plant glabrous, with creeping stems or rhizomes; upright stems 2 to 8 inches (5-20 cm) tall; flowers blue with a white-ringed yellow eye (Fig. 432) .... **BLUETS,** *Hedyotis caerulea* (L.) Hook.

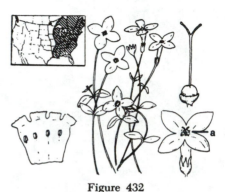

Figure 432

Synonym: *Houstonia caerulea* L.

A dainty little plant growing in moist grassy places in southeastern Canada and in most of the United States east of the Mississippi River. It ranges from 2 to 8 inches (5-20 cm) in height and often covers large spaces with its white or bluish white blossoms. The 4-lobed corolla is about 1/2 inch (1.3 cm) in diameter and has a yellow center. The basal leaves are obovate or spatulate 1/4 to 1/2 inch (0.6-1.2 cm) long and petiolate. Leaves on the stem are nearly sessile and more elongate. The plant is glabrous. It is also called Innocence and Quaker-ladies.

Small bee flies hover over the flowers on warm sunny afternoons and insert their long mouthparts into the narrow tubes for nectar.

THYME-LEAVED BLUETS, *Hedyotis michauxii* Fosberg (Synonym: *Houstonia serpyllifolia* Michx.), has creeping prostrate stems. It is found in moist soil in the mountains from Pennsylvania to Georgia and Tennessee. The 1/4 inch (0.6 cm) leaves have ovate to round blades abruptly narrowed at the base and short petioles. The blue flowers are on erect stems up to 4 inches (10 cm) tall.

4b Plants with minute hairs, without creeping stems; 1 to 4 inches (2.5-10 cm) tall; flowers lavender with a darker reddish eye (Fig. 433) .... **LEAST BLUETS** *Hedyotis crassifolia* Raf.

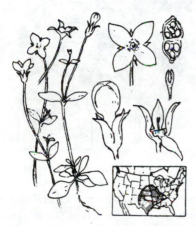

Figure 433

Synonym: *Houstonia minima* Beck; including *Houstonia patens* Ell.

A tiny plant that sometimes reaches a height of 5 inches (15 cm) but is usually about 2 inches (5 cm) high. It is common in dry soil. The small leaves are ovate to spatulate below and elliptic higher on the stem. The hairs on the leaves make them feel rough. The violet or purplish-blue flowers are single on a pedicel and about 1/3 inch (0.8 cm) broad. The eye of the flower is reddish purple with small purple lines radiating from the center.

5a Stem leaves ovate to lanceolate, rounded at the base (Fig. 434) ....
.... **LARGE HEDYOTIS,** *Hedyotis purpurea* (L.) Torr. & Gray

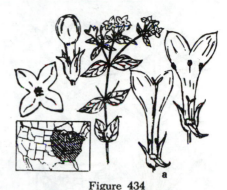

Figure 434

Synonym: *Houstonia purpurea* L., LARGE HOUSTONIA

This tall *Hedyotis* grows from 4 to 20 inches (10-50 cm) high, usually in rocky open woods and clearings. Purplish or lilac flowers (or nearly white) 3/8 inch (1 cm) long have a gradually flaring corolla and are in terminal cymose clusters. The leaves are rounded at the base, ovate to lanceolate, 3/4 to 2 inches long (2-5 cm) and 1/4 to 1 1/8 inch (0.5-3 cm) wide with 3 to 7 veins running lengthwise. ★ – NC

In species of this genus there are two types of flowers. In one the style is taller than the anthers (a); in the other the stamens are attached higher in the corolla tube and the style is short so that the stigma is below the anthers. This is an adaptation which allows both cross pollination (in the tall style flower) and self pollination.

5b Stem leaves linear to narrowly lanceolate, tapering at the base (Fig. 435) .... **SLENDER-LEAVED HEDYOTIS,** *Hedyotis nuttalliana* Fosberg

6a (1) Flowers white or greenish 7

6b Flowers yellow (Fig. 436) .... **YELLOW BEDSTRAW,** *Galium verum* L.

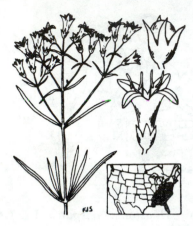

Figure 435

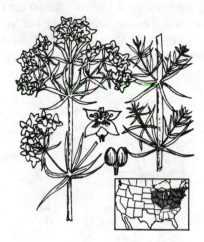

Figure 436

Synonym: *Houstonia tenuifolia* Nutt.

Dry soil. A much branched, slender, glabrous perennial attaining a height of 6 to 16 inches (15-40 cm). All except the basal leaves are very narrow and often more than an inch (2.5 cm) long. The purple funnel-shaped flowers about 1/4 inch (6 mm) long are on slender pedicels 1/4 inch to 5/8 inch (5-15 mm) long. The calyx lobes are shorter than the tube. The small leaves of the basal rosette may be still evident at flowering. They are ovate to elliptic and petiolate.

Another narrow-leaved Hedyotis, *H. longifolia* (Gaertn.) Hook., grows in dry forests and clearings over the United States east of the Rocky Mountains. This species is not as slender as the preceding. The pedicels of the purple to white 1/4 to 3/8 inch (6-9 mm) flowers are shorter than the corollas. In fruit the calyx lobes are longer than the capsule. In the narrow leaves of these two species there is only one main vein.

The stems are smooth or slightly roughened and achieve as much as 2 feet (61 cm) or more in length. The linear leaves measure 1/3 to 1 inch (0.8-2.5 cm) in length and are borne in 6's or 8's at each node. The flowers are bright yellow and crowded into dense showy inflorescences. The fruit is about 1/16 inch (1.6 mm) in diameter and is smooth. This plant is a native of Europe and Asia and has now spread in fields and roadsides where it blooms in spring and summer.

This species of *Galium* has been put to many uses. The plant is fragrant when dried and so made pleasant mattress stuffing. The leaves cause milk to curdle—a property that was used in cheese making. The roots yield a red dye.

7a Leaves ovate; flowers greenish (Fig. 437) .... **WILD LIQUORICE,** *Galium circaezans* Michx.

Figure 437

Found in rich dry woods, this plant is 8 to 24 inches (20-60 cm) tall with ovate or oval leaves in whorls of 4. The leaves have 3 to 5 veins and are 3/4 to 2 inches (2-5 cm) long by 1/4 to 1 inch (1-2.5 cm) wide. The greenish-yellow flowers are sessile in few-flowered cymes. The black fruit is covered with small bristles.

Galium triflorum Michx., SWEET-SCENTED BEDSTRAW, has elliptic leaves 1/4 to 3/8 inch (7-10 mm) wide with one main straight vein and a mucronate tip, borne usually 6 to a node. The small greenish-white flowers are usually borne 3 to an axillary or terminal peduncle. The common name comes from the vanilla-scented foliage. This plant is native around the Northern Hemisphere from cool climates south into Florida and Mexico.

7b Leaves linear to narrowly elliptic; flowers white **8**

8a Plant smooth, glabrous; leaves mostly in whorls of 4; inflorescences dense (Fig. 438) .... **NORTHERN BEDSTRAW,** *Galium boreale* L.

Figure 438

This bedstraw is an erect, smooth and glabrous plant 8 to 39 inches (20-100 cm) high, growing in rocky soil or along streams. The lanceolate or linear leaves, 1/2 to 2 inches (1.3-5 cm) long, are usually in whorls of 4. The white, 4-petaled flowers, 1/8 to 1/4 inch (3-6 mm) long, are in a dense terminal panicle. The fruit is usually rough when young, but often becomes smooth when it matures. This blooms from May to August.

Galium obtusum Bigelow also has white flowers and linear leaves in 4's. The leaves are 3/4 inch (2 cm) long or shorter. The most obvious difference between this and the Northern Bedstraw is the tiny flowers, less than 1/8 inch (3 mm) wide, borne in groups of only 2 to 4 in branching, diffuse inflorescences. This is a species of shady, moist soils from Quebec to South Dakota south to Florida and Texas.

SOUTHWESTERN BEDSTRAW, *Galium virgatum* Nutt., is a small hispid species of barren ground and the prairies. The leaves, less than 1/2 inch (1.3 cm) long, are borne in whorls of 4. The fruit is covered with barbed bristles; the tiny white flowers are single and sessile in the leaf axils. The small fruits are bristly. It is found from Tennessee to Louisiana and west to eastern Kansas and Texas.

8b **Plant rough, the leaves and stem with short, stiff bristles; leaves mostly in whorls of 6 to 8 (Fig. 439)** .... **CLEAVERS,** *Galium aparine* **L.**

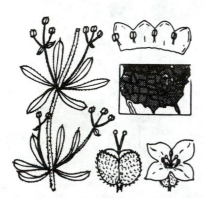

Figure 439

A weak-stemmed plant of woods and shady areas. The stems are 8 to 39 inches (20-100 cm) long and scramble over other plants, often forming tangled patches that are hard to walk through. The margins and midribs of the leaves and the stem have downward-pointing prickles. The whorled linear to oblanceolate leaves 3/4 to 1 3/4 inches (2-4.5 cm) long are in 6's or 8's. The tiny white flowers are 1 to 3 in axillary cymes. The bur-like fruit is densely covered with hooked prickles. Other names are Goose Grass and Bedstraw.

The roasted seeds make a coffee substitute. The old European herbals recommend the broth as an aid to staying slim, perhaps a successful one in some ways because as a medicine the plant was considered a diuretic and laxative.

Galium tinctorium L. which ranges throughout North America is also a prickly, "cleaving," straggling plant. Its leaves are in whorls of 5 or 6. The corollas are only 3 lobed. Its name means that it has been used as a dye plant.

CAPRIFOLIACEAE

HONEYSUCKLE FAMILY

(Fig. 440) .... **HORSE GENTIAN,** *Triosteum perfoliatum* **L.**

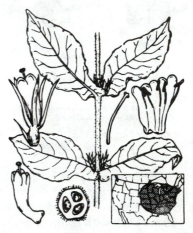

Figure 440

This species is a stout coarse plant, 2 to 5 feet (61-150 cm) high, of rich soil found from Massachusetts and Nebraska south to Alabama. The stem and the ovate or oval, connate-perfoliate leaves are glandular pubescent. The purplish-brown flowers, 1/2 to 3/4 inch (1.3-1.9 cm) long, are borne in the leaf axils. The corolla is slightly swollen at one side near the base. The 5 sepals are linear and nearly as long as the corolla. The fruit is a dull orange drupe. It is also called Tinker's Weed.

The "berries" have been dried and roasted for use as a coffee substitute and the leaves have been used as a purge to treat fevers.

SCARLET-FRUITED HORSE GENTIAN, *Triosteum aurantiacum* Bicknell, is a similar plant with purplish-red flowers, sessile hardly-clasping leaves, and orange-red fruits.

CAMPANULACEAE

BLUEBELL FAMILY

1a Leaves all round to ovate and those on the stem cordate-clasping; flowers sessile or nearly so in the axils of upper leaves (Fig. 441) .... **VENUS'S LOOKING-GLASS,** *Triodanis perfoliata* (L.) Nieuwl.

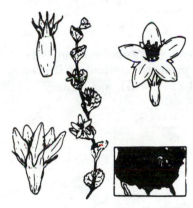

Figure 441

Synonym: *Specularia perfoliata* (L.) DC.

A common plant of dry woods and roadsides. The stem 4 to 39 inches (10-100 cm) high has many cordate-clasping, toothed leaves. The violet-blue flowers, 1/2 to 3/4 inch (1.3-1.9 cm) broad, are borne in the leaf axils, either singly or in 2's or 3's. The corollas are regular and the lobes spread wide from a short tube. The plant is erect or branched from the base with the branches slightly spreading and then ascending.

Triodanis leptocarpa (Nutt.) Nieuwl. is a species with purple flowers native from southern Minnesota to Montana and south to Texas and Arkansas. It differs most obviously from *T. perfoliata* in the leaves, which are oblanceolate or elliptic low down on the stem and become narrower higher until the leaves subtending the flowers are linear. Both of these

species have smaller cleistogamous flowers lower on the stem.

1b Leaves of stem narrowly lanceolate to linear, basal leaves long petiolate and round to ovate; flowers on long pedicels (Fig. 442) .... **HAREBELL,** *Campanula rotundifolia* L.

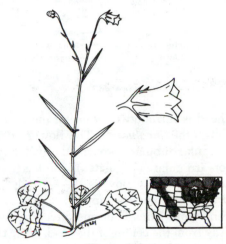

Figure 442

Well drained meadows and woods, cliffs. Found worldwide in the northern latitudes and in the mountains, this hardy but fragile-looking plant with narrow stem leaves and thin pedicels produces large nodding bell-shaped blue flowers. The corollas are 1/2 to 1 1/8 inch (1.5-3 cm) long with short lobes. The calyx lobes are long and very narrow, shorter than the tube and spreading in flower. At the base of the stem is a rosette of broadly ovate or cordate-rounded leaves on long petioles, the blades up to 3/4 inch (2 cm) long with toothed margins. These may gradually grade into the linear stem-type leaves.

COMPOSITAE [ASTERACEAE]

COMPOSITE FAMILY

In this family each head is a composite blossom consisting of many individual flowers (Fig. 443).

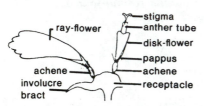

Figure 443

Each head is surrounded by one or more rings of bracts (the *involucre*). The flowers have either regular tubular corollas (the *disk* flowers) or irregular ("ligulate") corollas (*ray* flowers). In the ray flowers the 5 petals are united and the tongue-shaped petal-like portion bends and spreads in one direction, usually away from the center of the head. Since the flowers are crowded and protected by the involucre, the sepals do not have a protective function as in most buds, but are modified into bristles, hairs, or scales which often assist seed dispersal and are called collectively the *pappus*. Since the ovaries are inferior, the pappus is found at the top of the ovary. The flowers in a head are inserted on an expanded tip of the flower stalk called the *receptacle*. The involucre surrounds the receptacle. In some species the individual flowers are accompanied by a bract also inserted on the receptacle, in which case the receptacle is called chaffy. When the bracts are absent the receptacle is called naked.

1a Flowers all with tongue-shaped corollas (ray flowers) (Fig. 444); sap milky or colored 2

Figure 444

1b Flowers with tubular corollas (disk flowers) present; ray flowers, if present, marginal (Fig. 445); sap watery 7

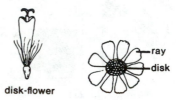

Figure 445

2a Pappus of simple unbranched bristles .. 3

2b Pappus of scales and bristles (genus KRIGIA) 5

3a Leaves all basal; heads solitary on naked scapes 4

3b Leaves basal and at least a few (sometimes much reduced) on the stem; heads several (Fig. 446) .... **LEAFY-STEMMED FALSE DANDELION,** *Pyrrhopappus carolinianus* (Walt.) DC.

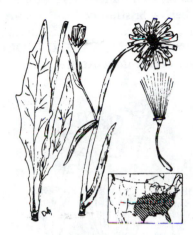

Figure 446

4a Heads with many flowers; leaves with pinnately toothed or lobed margins (Fig. 447a) .... **DANDELION,** *Taraxacum officinale* Weber

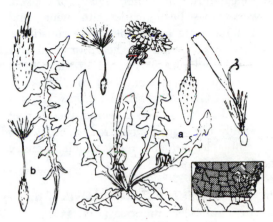

Figure 447

Dry fields. The yellow heads of this glabrous plant 8 to 39 inches (20-100 cm) high have a diameter of 1 to 1½ inches (2.5-4 cm). The entire or pinnately dissected leaves have a length of 3 to 10 inches (7.5-25 cm) and a width ranging up to 2 3/8 inches (6 cm). Several terminal heads may be borne on one branching stem. The involucre is 3/8 to 1 inch (1-2.5 cm) long. The hairy white ring at the base of the pappus is characteristic.

ROUGH FALSE DANDELION, *Pyrrhopappus grandiflorus* Nutt., is found on prairies from Kansas to Texas. The scape bearing the flower head is naked or with 1 to 2 small bracts. There may be a number of scapes from the rosette. The pinnately lobed leaves are pubescent on the margins. Buried below the long vertical rhizome is a tuberous root.

This is probably one of the most common and widely distributed plants of the world. Anyone who has tried to keep it out of a lawn knows how stubbornly and successfully it resists eradication. It is originally a native of Eurasia, where there are a number of other species of *Taraxacum* as well. The plant usually grows 2 to 20 inches (5-51 cm) high.

The leaves are pubescent on the midrib or glabrous, variously lobed or toothed, with a large terminal lobe. The inner bracts of the involucre are longer than the outer and erect in flower but reflexed in fruit. The shorter outer bracts are always reflexed. The achenes are gray- or olive-brown and rough. The pappus is white. Bees, flies, moths and butterflies visit the flowers.

RED SEEDED DANDELION, *T. laevigatum* (Willd.) DC., (Fig. 447b) (Synonym: *T. erythrospermum* Andrz.) has also been introduced from Europe and is now found in much of Canada and the United States. The achenes of this species are a beautiful reddish brown with sharp spines above. The leaves are lobed over their whole length and the lobes are deep and narrow.

Dandelions, although weeds, are also useful. The leaves gathered before flowering can be eaten raw in "wilted" salads, or cooked. Flowers have been made into wine, and roasted roots are a current coffee substitute. Root tea was a remedy for heartburn and liver disease. Dried root was listed in the official U.S. Pharmacopoeia from 1831 to 1926.

CHICORY, *Cichorium intybus* L., is a summer blooming blue-flowered member of the family often seen along roadsides. It also has edible leaves, and roots which intensify the flavor of coffee.

4b   Heads with few flowers; leaves with entire, pubescent margins (Fig. 448) .... PRAIRIE FALSE DANDELION, *Nothocalais cuspidata* (Pursh) Greene

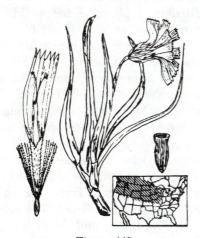

Figure 448

Synonym: *Agoseris cuspidata* (Pursh) Raf.

Stony dry soil or moist meadows. The basal leaves are linear, 3 to 8 inches (7.5-20 cm) long, less than 1/2 inch (1.3 cm) wide, and very pubescent on the margins. The scape is thick and becomes extremely densely hairy at the top. Each rosette may produce several

scapes. The flowering head is about 2 inches (5 cm) broad, the yellow rays are notched at the tip, and the achenes are smooth.

5a   Bracts of the involucre 8 or less; pappus scales 5 (to 8) alternating with an equal number of bristles, or bristles none (Fig. 449) .... WESTERN DWARF DANDELION, *Krigia occidentalis* Nutt.

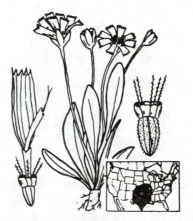

Figure 449

Prairies. Leaves basal, oblanceolate, entire or more rarely pinnately dissected, about half as long as the 2 to 8 inch (5-20 cm) glandular-hairy scapes. The bracts are 5 to 8, erect; the rays are yellow and 5-toothed. The pappus scales are usually 5, sometimes up to 8, and alternate with an equal number of bristles or with no bristles at all.

5b   Bracts of the involucre 9 to 18 6

6a Pappus bristles **5**, alternating with an equal number of scales (Fig. 450) .... **CAROLINA DWARF DANDELION,** *Krigia virginica* (L.) Willd.

6b Pappus bristles **many** (25 to 40); pappus scales very short, about 10 (Fig. 451) .... **DWARF DANDELION,** *Krigia dandelion* (L.) Nutt.

Figure 450

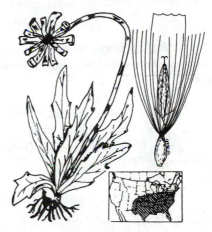

Figure 451

A common plant of dry sandy soil. This species, 2 to 16 inches (5-40 cm) high, is one of several Compositae that resemble the common Dandelion of roadsides and lawns. However, the yellow heads about 1/2 inch (1.3 cm) broad are smaller than those of the Dandelion. The flowering stems are usually 3 to 10 from a rosette, although there is only one in small plants and sometimes up to 20 or more. The leaves are linear to oblanceolate or obovate, from 1/2 to 4 5/8 inches (1.5-12 cm) long and 1/4 to 1 1/8 inch (0.5-3 cm) wide. They may be entire or pinnately lobed. The 5 thin scales of the pappus alternate with 5 long bristles (a).

Moist, open soil. The slender leafless scape rises to a height of 4 to 20 inches (10-50 cm) and is usually solitary. The entire or few-lobed leaves are all basal, 3/8 to 8 inches (2-20 cm) long, and very narrow or up to 1 inch (2.5 cm) broad. The golden-yellow heads are about 1 inch (2.5 cm) across, with long flowers and shorter involucral bracts 3/8 to 5/8 inch (0.9-1.4 cm) long. The scales are inconspicuous compared to the 25 to 40 pappus bristles. Another name is Potato Dandelion, in reference to the small tubers produced by the underground stems.

VIRGINIA GOATSBEARD, *Krigia biflora* (Walt.) Blake, produces 1 to 6 flowering heads on an 8 to 32 inch (20-80 cm) stem which bears a few sessile, often reduced, leaves. The deep orange flowers have a pappus of from 20 to 35 bristles. It is found in fields and roadsides from Massachusetts to Virginia and in the mountains to Georgia and west to Manitoba, Colorado and Arizona.

7a (1) Heads with disk flowers (with tubular corollas) only 8

7b Heads with both disk and ray flowers, the ray flowers marginal 11

8a Heads white to gray or pinkish; leaves usually pubescent below; stem leaves reduced 9

8b Heads yellow; leaves sparsely hairy or glabrous; stems leafy (Fig. 452) .... COMMON GROUNDSEL, *Senecio vulgaris* L.

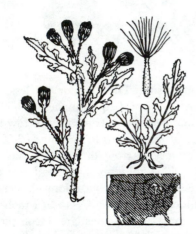

Figure 452

This is an annual weed which came from Europe and is widely distributed. It begins blooming early and continues throughout the growing season. Its hollow stems attain a height of 4 to 20 inches (10-51 cm). The leaves are pinnately lobed and coarsely toothed and occur the whole length of the stem. The stems are much branched with one to several heads on smaller branches at the tips. The yellow heads of disk flowers are 3/8 inch (1 cm) wide or half that. The involucral bracts of the heads are black tipped. The pappus has many bristles as long as or longer than the corolla.

The plant contains alkaloids which have killed livestock that ate them in quantity.

9a Basal leaves lobed .... SWEET COLTSFOOT, *Petasites frigidus* (L.) Fries (see Fig. 468)

This species has 2 types of heads, one with disk flowers only and one with disk and ray flowers. Since it is more closely related to other species with ray flowers it is described in that section of the key.

9b Basal leaves entire; heads with red styles 10

10a Basal and terminal leaves 3 to 5 veined, the blade broad, ovate or obovate, more than 1/2 inch (1.5 cm) wide (Fig. 453) .... PLANTAIN-LEAVED PUSSY-TOES, *Antennaria plantaginifolia* (L.) Richards

Figure 453

A white-woolly plant growing in the dry soil of open woods. The plants are either male or female. The heads are borne in clusters at the tips of stems 4 to 16 inches (10-40 cm) tall carrying reduced linear or lanceolate leaves. The heads of male flowers are smaller than the female, with white tips on the bracts, and tubular flowers with a thin pappus. The female flowers are softly hairy with a dense pappus which obscures the corollas and have a 2-branched, often red, style. The basal leaves and the ones at the end of the creeping stems are spatulate, with a narrow petiolar portion and an ovate or obovate blade with 3 to 5 veins. The larger leaves are 3/4 to 2¼ inches (2-6 cm) long and 1/2 to 2 inches (1.5-5 cm) wide. Several varieties based on involucre size and leaf pubescence are recognized.

This has been one of the myriad of plants in eastern North America used in treatments of rattlesnake bites.

10b Basal and terminal leaves single veined, the blade elliptic or oblanceolate, gradually narrowed to the petiole, 1/2 inch (1.5 cm) or less wide (Fig. 454) .... SMALLER CAT'S-FOOT, *Antennaria neglecta* **Greene**

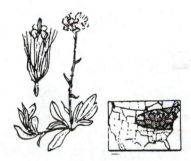

Figure 454

Dry, open woods. This plant is very similar to *Antennaria plantaginifolia*. It has the same wooliness of foliage and separate male and female plants. In this species the flowering stems are 1½ to 14 inches (4-35 cm) tall. The basal leaves and ones at the ends of creeping stems have a narrower blade, mostly less than 1/2 inch (1.5 cm) wide. The blade is elliptic or oblanceolate, and narrows gradually to the pe-

tiole. It is single-veined or with 3 faint veins. This species also has a number of varieties. Other names for members of this genus are Everlasting, and Ladies'-tobacco.

11a (7) Ray flowers yellow 12

11b Ray flowers white, purplish or pink 19

12a Basal leaves not present at flowering time; leaves all basal; scape with long scaly bracts; rays in several rows (Fig. 455) .... COLTSFOOT, *Tussilago farfara* **L.**

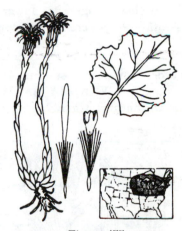

Figure 455

This European plant now blooms in early spring along roadsides and in other disturbed soil in northeastern North America. The heads of yellow flowers on many-bracted stems 2 to 20 inches (5-50 cm) tall bloom before the basal leaves develop. The heads have many flowers with narrow rays arranged in several rings and when in full bloom are up to 1 1/8 inch (3 cm) wide. The pappus bristles are numerous so that a fruiting head is round like that of a dandelion. The leaves, arising from the rhizome, are cordate or nearly round, scalloped with gentle lobes and toothed margins, white tomentose beneath and green above. They measure 2 to 8 inches (5-20 cm) across (a sizeable colt's track).

The plant is widely known for its use in respiratory afflictions. Smoke from the leaves is inhaled for asthma. Leaf extract and sugar make cough syrup, cough drops (and candy). Dry leaf tea is drunk for colds.

12b Basal leaves present at flowering; leaves both basal and on the stem **13**

13a Stems with leaves alternate; lower stem leaves sessile and pinnately lobed or toothed **14**

13b Stem with leaves opposite; lower stem leaves entire or with crenate margins .. **17**

14a Plant woolly or tomentose (Fig. 456) ....
.... **WOOLLY RAGWORT,**
Senecio tomentosus **Michx.**

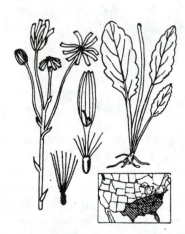

Figure 456

Open places, especially sandy soil. This perennial with tomentose stems grows 8 to 28 inches (20-70 cm) tall. The lower leaves are on long petioles, the shallowly toothed blades are lanceolate to elliptic, some as long as 8 inches

(20 cm). The stem leaves become sessile and may be slightly toothed. There are usually a number of heads with 10 to 15 yellow rays 3/8 inches (1 cm) long or less.

PRAIRIE RAGWORT, S. *plattensis* Nutt., is similar. It has smaller basal leaves and the stem leaves are deeply pinnately cut. It ranges mostly in the plains from Montana and Arizona east to a few sites in Virginia and Tennessee.

14b Plants glabrous or largely so **15**

15a Basal leaves cordate, broad, on long narrow petioles (Fig. 457) ....
.... **GOLDEN RAGWORT,**
Senecio aureus **L.**

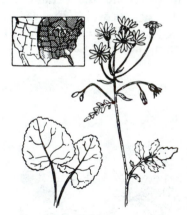

Figure 457

Golden Ragwort is an attractive yellow-flowered member of the Compositae that blooms from May to July. It grows mostly in wet open places. The daisy-like flower heads are about 3/4 inch (2 cm) broad with 8 to 12 ray-flowers in each head. The sessile stem leaves are often cut and pinnatifid; the long petioled basal leaves are cordate-ovate with toothed edges. The entire plant grows from 9 to 30 inches (23-76 cm) high. This plant was once listed in the National Formulary for pharmaceutical use and has been used by eastern Indians. It was used for pain and hemorrhage in labor, as a stimulant and a diuretic.

15b Basal leaves obovate to elliptical or nearly round, blades tapering to the petiole .. 16

16a Lower leaves mostly oblong to oblanceolate or elliptic; plant occasionally with short stolons (Fig. 458) .... BALSAM GROUNDSEL, *Senecio pauperculus* **Michx.**

Figure 458

Moist soil. This 4 to 20 inch (10-51 cm) tall plant bears comparatively few flowering heads. The petioles of the basal leaves are relatively short, the blade and petiole together are usually less than 4 3/4 inches (12 cm). The stem leaves are usually lyrate and sessile. The upper parts of the plant are glabrous but the basal parts are frequently somewhat tomentose.

Several species of cultivated plants fall in this genus. DUSTY-MILLER is grown for its white-tomentose leaves. CINERARIA is a well-known florists' plant available with flower heads in all colors except yellow.

16b Basal leaves obovate to orbicular; plant with long slender stolons or rhizomes (Fig. 459) .... ROUND-LEAF SQUAW-WEED, *Senecio obovatus* **Muhl. ex Willd.**

Figure 459

Moist woods and open places. This usually glabrous perennial may attain a height of 2 feet (61 cm) with rather thick basal leaves, with round or obovate blades narrowed to the petiole, the blades up to 2¼ inches (6 cm) long. The lower stem leaves are often deeply pinnately lobed and the lobes toothed. The yellow rays are occasionally few and short but usually make the heads conspicuous.

17a Leaf margins crenate; rays 5 (Fig. 460)
.... **CHRYSOGONUM,**
Chrysogonum virginianum L.

Figure 460

This little, dry soil, pubescent or hirsute perennial grows to a height of 3 to 20 inches (7.5-50 cm). The thin leaves with ovate blades and long petioles may attain a length of 4 inches (10 cm) and a width of 2¼ (6 cm). The yellow rays about 1/2 inch long (1-1.5 cm) are usually 5 but may vary by one or two in either direction. The bracts of the involucre are in 2 sets of about 5. The inner set is small but the outer set is green and leafy, up to 3/8 inch (1 cm) long. The plant often flowers when it is quite small, sometimes appearing no more than a rosette of leaves with a flower nestled in the center. Another name is Green-and-gold. A southern variety, var. *australe* (Alexander) Ahles, is low growing, less than 6 inches tall, and produces stolons.

17b Leaf margins entire; rays usually 8 to 15
.... **18**

18a Leaves elliptic to ovate; rays 10 to 15; plant glandular hirsute (Fig. 461) ....
.... **LEOPARD'S-BANE,**
Arnica acaulis (Walt.) B.S.P.

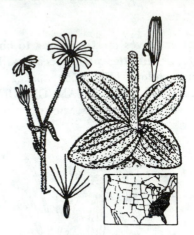

Figure 461

This pubescent plant bearing several heads of yellow flowers grows to a height of 8 to 32 inches (20-80 cm). The stem usually bears 2 or 3 pairs of opposite, sessile leaves and has a rosette of broad, usually elliptic or ovate leaves with 5 to 7 parallel veins at its base. The heads have 10 to 15 ray flowers 1/2 to 1 1/8 inch (1.5-3 cm) long. The pappus is yellowish white.

18b Leaves lanceolate; rays 6 to 10, usually 8; plants glabrous or pubescent (Fig. 462) .... LANCE-LEAVED TICKSEED, *Coreopsis lanceolata* L.

Figure 462

An attractive plant of rich soil found in most of the eastern half of the United States. It is native in its western range but has escaped from cultivation in the East. The plant grows 1/3 to 3 feet (10-91 cm) high with a slender stem and lanceolate leaves. The showy flower heads, 1½ to 3 inches (3.8-7.5 cm) broad, have 6 to 10 bright yellow ray-flowers and many disk-flowers. The achenes have a wing and 2 short teeth. The involucral bracts are up to 3/8 inch (1 cm) long and lanceolate or ovate.

This genus contains many species. In the southeastern United States alone there are 19, most of them blooming in late summer and fall. Some species are cultivated, most have bright yellow rays but a few are rose purple.

19a (11) Heads small, the disk 1/4 inch (6 mm) broad or less **20**

19b Heads with the disk more than 1/4 inch (6 mm) broad **21**

20a Leaves opposite, entire (Fig. 463) .... GALINSOGA, *Galinsoga quadriradiata* Ruiz & Pavon

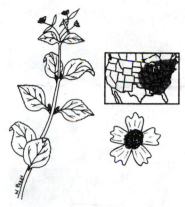

Figure 463

Synonym: *Galinsoga ciliata* (Raf.) Blake

This hairy annual weed from Central and South America is naturalized throughout North America and in the Old World. It blooms from spring until fall, and into winter in the far south. The stems are much branched, erect or somewhat spreading, to 28 inches (10-70 cm) tall. The leaves are opposite and petiolate; the leaf blades are ovate, 1 to 4 inches (2.5-10 cm) long with coarse teeth. The heads are numerous and small, the disk 1/4 inch (6 mm) wide or less, surounded by 3 to 6 (usually 5) white (rarely pink) rays less than 1/8 inch (3 mm) long.

QUICKWEED, *Galinsoga parviflora* Cav., is less common but worldwide and also flowers in spring. It is less hairy than *G. quadriradiata* and the ray flowers lack a pappus. The leaves tend to be lanceolate-ovate. This plant has been cooked as a green vegetable.

20b Leaves alternate, pinnately several times dissected into narrow segments (Fig. 464) .... YARROW, *Achillea millefolium* L.

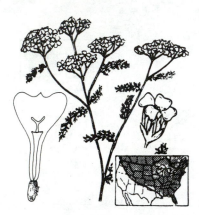

Figure 464

22a Leaves entire (Fig. 465) .... WESTERN DAISY, *Astranthium integrifolium* (Michx.) Nutt.

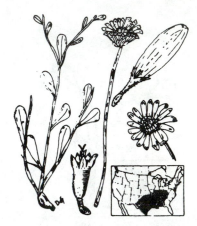

Figure 465

Yarrow is a familiar and common plant 1 to 4 feet (30-122 cm) high, of fields and roadsides found in the eastern half of North America and in Europe and Asia. The leaves are finely dissected into narrow segments. The numerous aromatic heads are on stiff peduncles and form dense, nearly flat clusters. The white rays are 3 to 6 in number. Occasionally the flowers are rose-pink. This plant is sometimes cultivated and is also called Milfoil. In its vegatative stage it is often mistaken for a fern.

Eaten, this plant causes gastric distress. As an aromatic poultice or vapor it is a cold cure, as tea it is considered a tonic, stimulant and diuretic, and as a shampoo or hair rinse, it supposedly treats baldness and brightens blonde hair.

21a Pappus none or only a very short ring on the achene **22**

21b Pappus of long bristles **24**

This slender plant grows in moist or dry soil, in woods or open ground. The leafy stems are single or branched, 3 to 18 inches (8-45 cm) tall, and end in a long peduncle with a single daisy-like head. The entire-margined alternate leaves are oblanceolate or spatulate low on the stem and become elliptic near the top, the longest about 3 inches (8 cm) long and 3/4 inch (2 cm) wide. The disk flowers are yellow and the 8 to 22 ray flowers, 3/8 inch (1 cm) long or half that, are blue, purple or white.

ENGLISH DAISY, *Bellis perennis* L., is a low cultivated daisy which has run wild into lawns and waste places in the northern United States, in the mountains to North Carolina, and the western Cascades. Another, appropriate, name is Lawn Daisy. The leafless scapes each bearing a single head, 1 to 2 inches (2.5-5 cm) across rise to a height of 3/4 to 6 inches (1.9-15 cm) from a rosette of pubescent, spatulate leaves. The numerous rays may be rose, purple or white.

22b Leaves lobed, dissected or toothed **23**

23a Leaves several times dissected into narrow final segments (Fig. 466) .... DOG FENNEL, *Anthemis cotula* L.

23b Leaves only once pinnately dissected or toothed (Fig. 467) ...... OX-EYE DAISY, *Leucanthemum vulgare* Lam.

Figure 466

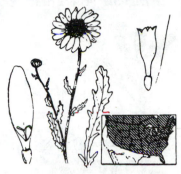

Figure 467

Synonym: *Chrysanthemum leucanthemum* L.

This branched pungent herb 4 to 24 inches (10-60 cm) tall has alternate leaves that are pinnately dissected 2 to 3 times into narrow segments. The heads are terminal, either solitary on long branches or on shorter branches near the top of the main stem. They are daisy-like, with yellow disk flowers and a ring of 10 to 16 white ray flowers 1/4 to nearly 1/2 inch (5-11 mm) long. The plant grows in lawns and waste ground, having been naturalized from Europe. The common name comes from the fennel-like leaves and the rather unpleasant odor. *Anthemis arvensis* L. is similar but hairy and without the pungent odor.

An attractive and abundant plant of fields and waste places found in most of North America. It was introduced from Europe and is so well established in certain areas as to be a very troublesome weed. It grows to 3 feet (91 cm) high with large showy heads 1 to 3 inches (2.5-7.5 cm) broad, and deeply-cut attractive leaves. The disk is yellow and is bordered by 15 to 35 pure white ray-flowers. Another name is Marguerite.

24a Stems with leaves reduced to elongated scales; blade-bearing leaves all basal, very different from the stem scales (Fig. 468) .... **SWEET COLTSFOOT,** *Petasites frigidus* (L.) Fries

Figure 468

Heads in this species are of 2 kinds, the "female" heads with nearly all pistillate flowers which produce fruits and "male" heads which have both stamens and pistils but the pistils not producing fruit. The flowers are white, those in the male heads with tubular corollas and those in the female with short rays. The fragrant flowers are in clusters or racemes at the tops of bracteate stems 4 to 20 inches (10-50 cm) tall. The basal leaves arise from a rhizome. The blades, on long petioles, are very hairy underneath, deeply lobed and large, from 2 to 10 inches (5-25 cm) wide. It grows in swamps and streams the world around in the far north. Three varieties are recognized. Most of the plants that reach south into the United States are var. *palmatus* (Ait.) Cronq. In this variety the leaves are palmately veined and palmately lobed at least 2/3 of the way to the base.

ARROW-LEAF SWEET COLTSFOOT, *P. sagittatus* (Banks ex Pursh) Gray, is a more northern plant that extends south into Minnesota, South Dakota and Colorado. It is much like *Petasites frigidus* except that the leaves are only toothed or nearly lobed.

BUTTERBUR, or Butterfly Dock, *P. hybridus* (L.) Gaertn., Mey. & Scherb., is an introduced European weed which has escaped from cultivation and grows from Massachusetts to Delaware. The leaves are cordate and the rayless flowers purple.

24b Stems with blade-bearing, alternate leaves gradually differing from the basal leaves **25**

25a Heads 1 to 2 inches (2.5-5 cm) broad; ray flowers 50 to 100; rays about 1/16 inch (1-2 mm) or more broad; plant with stolons (Fig. 469) .... **ROBIN'S PLANTAIN,** *Erigeron pulchellus* Michx.

Figure 469

Banks and wooded hillsides. The plant has softly hairy stems from 6 to 24 inches (15-60 cm) tall. The basal leaves are oblanceolate to obovate and often toothed on the upper 1/2 to 1/3. The upper leaves become lanceolate and smaller upwards. At least the margins of the leaves have long hairs. The heads are single or commonly 2 to 4 on peduncles at the top of the stem, 1 to 2 inches (2.5-5 cm) broad.

The ray flowers 1/2 to 3/4 inch (1-2 cm) broad number about 50 (to 100) and are blue or sometimes pink or white. The pappus is tan. ★ — MN

25b Heads 1/2 to 1 inch (1.3-2.5 cm) broad; ray flowers numerous (150-400); rays very narrow (0.5 mm); plant without stolons (Fig. 470) .... DAISY FLEABANE, *Erigeron philadelphicus* **L.**

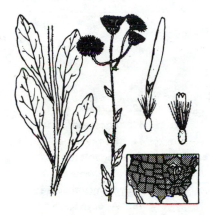

Figure 470

Daisy Fleabane is a common plant of fields and woods found in most of Canada and in al-most all of the United States. It grows 1 to 3 feet (30-91 cm) high with a slender, often branching stem, basal spatulate leaves, and clasping stem leaves. The heads are 1/2 to 1 inch (1.3-2.5 cm) broad and have about 150 to 400 light purple or pink ray-flowers. Another name is Philadelphia Fleabane. Oil extracted from the leaves and flowers has been used to hasten childbirth.

Erigeron is a common Composite genus of 149 species in North America that blooms in spring and early summer. Eight species occur in the Southeast and 57 species in the Northwest.

List of Families
In Phylogenetic Order

The treatments of families are arranged in this sequence.

Monocotyledonae

1. Alismataceae
2. Araceae
3. Commelinaceae
4. Bromeliaceae
5. Pontederiaceae
6. Liliaceae
7. Amaryllidaceae
8. Agavaceae
9. Iridaceae
10. Orchidaceae

Dicotyledonae

11. Santalaceae
12. Aristolochiaceae
13. Polygonaceae
14. Nyctaginaceae
15. Aizoaceae
16. Portulacaceae
17. Caryophyllaceae
18. Nymphaeaceae
19. Ranunculaceae
20. Berberidaceae
21. Papaveraceae
22. Fumariaceae
23. Cruciferae [Brassicaceae]
24. Sarraceniaceae
25. Crassulaceae
26. Saxifragaceae
27. Rosaceae
28. Leguminosae [Fabaceae]
29. Oxalidaceae
30. Geraniaceae
31. Polygalaceae
32. Euphorbiaceae
33. Buxaceae
34. Limnanthaceae
35. Malvaceae
36. Cistaceae
37. Violaceae
38. Passifloraceae
39. Onagraceae
40. Araliaceae
41. Umbelliferae [Apiaceae]
42. Cornaceae
43. Ericaceae
44. Primulaceae
45. Loganiaceae
46. Gentianaceae
47. Menyanthaceae
48. Apocynaceae
49. Asclepiadaceae
50. Convolvulaceae
51. Polemoniaceae
52. Hydrophyllaceae
53. Boraginaceae
54. Verbenaceae
55. Labiatae [Lamiaceae]
56. Solanaceae
57. Scrophulariaceae
58. Orobanchaceae
59. Plantaginaceae
60. Rubiaceae
61. Caprifoliaceae
62. Campanulaceae
63. Compositae [Asteraceae]

List of
Families by Common Name

Amaryllis family *Amaryllidaceae*

Arum family *Araceae*

Barberry family *Berberidaceae*

Birthwort family *Aristolochia-ceae*

Bluebell family *Campanulaceae*

Borage family *Boraginaceae*

Box family *Buxaceae*

Broom-rape family *Orobancha-ceae*

Buckbean family *Menyantha-ceae*

Buckwheat family *Polygonaceae*

Buttercup family *Ranunculaceae*

Cabbage family *Cruciferae*

Carpet Weed family *Aizoaceae*

Century Plant family *Agavaceae*

Composite family *Compositae*

Dogbane family *Apocynaceae*

Dogwood family *Cornaceae*

Evening Primrose family *Ona-graceae*

False Mermaid family *Limnan-thaceae*

Figwort family *Scrophularia-ceae*

Four O'clock family *Nyctagina-ceae*

Fumitory family *Fumariaceae*

Gentian family *Gentianaceae*

Geranium family *Geraniaceae*

Ginseng family *Araliaceae*

Heath family *Ericaceae*

Honeysuckle family *Caprifolia-ceae*

Iris family *Iridaceae*

Legume family *Leguminosae*

Lily family *Liliaceae*

Logania family *Loganiaceae*

Madder family *Rubiaceae*

Mallow family *Malvaceae*

Milkweed family *Asclepiadaceae*

Milkwort family *Polygalaceae*

Mint family *Labiatae*

Morning-glory family *Convolvu-laceae*

Nightshade family *Solanaceae*

Orchid family *Orchidaceae*

Parsley family *Umbelliferae*

Passion Flower family *Passiflora-ceae*

Pickerel-weed family *Ponteder-iaceae*

Pineapple family *Bromeliaceae*

Pink family *Caryophyllaceae*

Pitcher-plant family *Sarracenia-ceae*

Plantain family *Plantaginaceae*

Polemonium family *Polemonia-ceae*

Poppy family *Papaveraceae*

Primrose family *Primulaceae*

Purslane family *Portulacaceae*

Rockrose family *Cistaceae*

Rose family *Rosaceae*

Sandalwood family *Santalaceae*

Saxifrage family *Saxifragaceae*

Spiderwort family *Commelina-ceae*

Spurge family *Euphorbiaceae*

Stonecrop family *Crassulaceae*

Vervain family *Verbenaceae*

Violet family *Violaceae*

Waterleaf family *Hydrophylla-ceae*

Waterlily family *Nymphaeaceae*

Water-plantain family *Alismata-ceae*

Wood Sorrel family *Oxalidaceae*

Alphabetic Index of Genera by Family

This index includes all genera either pictured or discussed in the text.

MONOCOTYLEDONAE
AGAVACEAE, Century Plant family, 55
    Yucca, 55
ALISMATACEAE, Water-plantain family, 30
    Sagittaria, 30
AMARYLLIDACEAE, Amaryllis family, 54
    Hypoxis, 54
    Zephyranthes, 54
ARACEAE, Arum family, 31
    Acorus, 31
    Arisaema, 33
    Calla, 32
    Peltandra, 32
    Symplocarpus, 31
BROMELIACEAE, Pineapple family, 36
    Tillandsia, 36
COMMELINACEAE, Spiderwort family, 34
    Commelina, 34
    Tradescantia, 34
IRIDACEAE, Iris family, 55
    Iris, 57
    Nemastylis, 55
    Sisyrinchium, 56
LILIACEAE, Lily family, 37
    Allium, 41
    Asparagus, 37
    Calochortus, 49
    Camassia, 43
    Chamaelirium, 37
    Clintonia, 42
    Convallaria, 40
    Disporum, 52
    Erythronium, 40
    Maianthemum, 38
    Medeola, 44
    Muscari, 39
    Nothoscordum, 42
    Ornithogalum, 43
    Polygonatum, 38
    Scilla, 43
    Smilacina, 50
    Streptopus, 53
    Trillium, 44
    Uvularia, 51
    Veratrum, 49

    Zigadenus, 48
ORCHIDACEAE, Orchid family, 60
    Amerorchis, 67
    Aplectrum, 69
    Arethusa, 66
    Calypso, 65
    Coeloglossum, 66
    Corallorhiza, 63
    Cypripedium, 60
    Galearis, 67
    Isotria, 64
    Liparis, 70
    Listera, 69
    Pogonia, 65
    Spiranthes, 68
PONTEDERIACEAE, Pickerelweed family, 36
    Eichornia, 36

DICOTYLEDONAE
AIZOACEAE, Carpet Weed family, 75
    Mollugo, 75
APOCYNACEAE, Dogbane family, 172
    Amsonia, 172
    Apocynum, 173
    Vinca, 172
ARALIACEAE, Ginseng family, 159
    Aralia, 160
    Panax, 159
ARISTOLOCHIACEAE, Birthwort family, 71
    Asarum, 71
ASCLEPIADACEAE, Milkweed family, 173
    Asclepias, 173
BERBERIDACEAE, Barberry family, 102
    Caulophyllum, 102
    Jeffersonia, 102
    Podophyllum, 103
BORAGINACEAE, Borage family, 185
    Buglossoides, 186
    Cynoglossum, 185
    Lithospermum, 187

    Mertensia, 185
    Myosotis, 187
BUXACEAE, Box family, 147
    Pachysandra, 147
CAMPANULACEAE, Bluebell family, 211
    Campanula, 211
    Triodanis, 211
CAPRIFOLIACEAE, Honeysuckle family, 210
    Triosteum, 210
CARYOPHYLLACEAE, Pink family, 77
    Agrostemma, 77
    Arenaria, 80
    Cerastium, 82
    Moehringia, 81
    Myosoton, 82
    Silene, 77
    Stellaria, 81
CISTACEAE, Rockrose family, 148
    Helianthemum, 148
COMPOSITAE, Composite family, 212
    Achillea, 222
    Antennaria, 216
    Anthemis, 223
    Arnica, 220
    Astranthium, 222
    Bellis, 222
    Chrysogonum, 220
    Cichorium, 214
    Coreopsis, 221
    Erigeron, 224
    Galinsoga, 221
    Krigia, 212, 214
    Leucanthemum, 223
    Nothocalais, 214
    Petasites, 216, 224
    Pyrrhopappus, 213
    Senecio, 216, 218
    Taraxacum, 213
    Tussilago, 217
CONVOLVULACEAE, Morning-glory family, 176
    Calystegia, 176
    Convolvulus, 176
CORNACEAE, Dogwood family, 166

    Cornus, 166
CRASSULACEAE, Stonecrop family, 123
    Sedum, 123
CRUCIFERAE, Cabbage family, 108
    Alliaria, 117
    Arabis, 118
    Barbarea, 110
    Brassica, 111
    Camelina, 108
    Capsella, 114
    Cardamine, 120
    Conringia, 108
    Dentaria, 113
    Draba, 112
    Erophila, 112
    Erysimum, 108
    Hesperis, 116
    Iodanthus, 117
    Lepidium, 115
    Nasturtium, 116
    Sinapis, 111
    Sisymbrium, 110
    Thlaspi, 115
ERICACEAE, Heath family, 166
    Epigaea, 167
    Gaultheria, 166
EUPHORBIACEAE, Spurge family, 145
    Euphorbia, 145
FUMARIACEAE, Fumitory family, 104
    Corydalis, 104
    Dicentra, 104
GENTIANACEAE, Gentian family, 171
    Obolaria, 171
    Sabatia, 171
GERANIACEAE, Geranium family, 142
    Erodium, 142
    Geranium, 143
HYDROPHYLLACEAE, Waterleaf family, 181
    Ellisia, 182
    Hydrophyllum, 182
    Nemophila, 181
    Phacelia, 182

LABIATAE, Mint family, 190
  Glechoma, 192
  Hedeoma, 190
  Lamium, 193
  Leonurus, 192
  Prunella, 190
  Salvia, 190
  Scutellaria, 190
LEGUMINOSAE, Legume family, 132
  Astragalus, 135
  Baptisia, 137
  Lathyrus, 133
  Lupinus, 136
  Medicago, 138
  Melilotus, 137
  Psoralea, 136
  Tephrosia, 134
  Trifolium, 138
  Vicia, 132
LIMNANTHACEAE, False Mermaid family, 147
  Floerkea, 147
LOGANIACEAE, Logania family, 170
  Spigelia, 170
MALVACEAE, Mallow family, 148
  Malva, 148
MENYANTHACEAE, Buckbean family, 171
  Menyanthes, 171
NYCTAGINACEAE, Four O'Clock family, 75
  Mirabilis, 75
NYMPHAEACEAE, Waterlily family, 84
  Nelumbo, 85
  Nuphar, 84
  Nymphaea, 85
ONAGRACEAE, Evening Primrose family, 158
  Calylophus, 158
  Oenothera, 159
OROBANCHACEAE, Broomrape family, 202
  Conopholis, 203
  Orobanche, 202
OXALIDACEAE, Wood Sorrel family, 141
  Oxalis, 141
PAPAVERACEAE, Poppy family, 103
  Chelidonium, 104
  Sanguinaria, 103
  Stylophorum, 104
PASSIFLORACEAE, Passion Flower family, 158
  Passiflora, 158
PLANTAGINACEAE, Plantain family, 203
  Plantago, 203
POLEMONIACEAE, Polemonium family, 177
  Phlox, 177
  Polemonium, 177
POLYGALACEAE, Milkwort family, 144
  Polygala, 144
POLYGONACEAE, Buckwheat family, 72
  Polygonum, 72
  Rumex, 73
PORTULACACEAE, Purslane family, 76
  Claytonia, 76
PRIMULACEAE, Primrose family, 167

Anagallis, 169
Androsace, 168
Dodecatheon, 167
Lysimachia, 169
Trientalis, 168
RANUNCULACEAE, Buttercup family, 85
  Actaea, 94
  Anemone, 98
  Aquilegia, 87
  Caltha, 88
  Coptis, 101
  Delphinium, 86
  Hepatica, 97
  Hydrastis, 100
  Isopyrum, 100
  Myosurus, 88
  Pulsatilla, 97
  Ranunculus, 88
  Thalictrum, 95
  Trollius, 101
ROSACEAE, Rose family, 126
  Duchesnea, 127
  Fragaria, 126
  Geum, 127
  Potentilla, 127
  Waldsteinia, 127
RUBIACEAE, Madder family, 205
  Galium, 208
  Hedyotis, 206
  Mitchella, 206
SANTALACEAE, Sandalwood family, 71
  Comandra, 71
SARRACENIACEAE, Pitcher-Plant family, 122
  Sarracenia, 122

SAXIFRAGACEAE, Saxifrage family, 123
  Chrysosplenium, 123
  Heuchera, 124
  Mitella, 124
  Saxifraga, 125
  Tiarella, 125
SCROPHULARIACEAE, Figwort family, 195
  Castilleja, 195
  Collinsia, 200
  Gratiola, 196
  Linaria, 195
  Pedicularis, 196
  Penstemon, 200
  Scrophularia, 200
  Veronica, 196
SOLANACEAE, Nightshade family, 194
  Physalis, 194
  Solanum, 194
UMBELLIFERAE, Parsley family, 160
  Chaerophyllum, 162
  Erigenia, 163
  Heracleum, 161
  Osmorhiza, 162
  Polytaenia, 164
  Sanicula, 161
  Taenidia, 164
  Thaspium, 165
  Zizia, 163, 165
VERBENACEAE, Vervain family, 189
  Verbena, 189
VIOLACEAE, Violet family, 149
  Hybanthus, 149
  Viola, 149

Index and Pictured Glossary

(Glossary terms are printed in capital letters, common names are in roman type; scientific names in italics, family names in caps and small caps type, and names of plants mentioned incidentally or in synonomy are in parenthesis.)

A

ACCESSORY FRUIT: a fruit composed of other parts of the inflorescence in addition to the ovary and seeds.

ACHENE: a small, dry, indehiscent 1-seeded fruit with a thin wall. Fig. 11.

Achillea
- ☐ *millefolium*, 222

Acorus
- ☐ *americanus*, 31

Actaea
- (*alba*), 95
- ☐ *pachypoda*, 94, 95
- ☐ *rubra*, 94

ACTINOMORPHIC: radially symmetrical; regular; capable of being divided into equal halves along several planes. Fig. 471

Fig. 471

ACUMINATE: with a sharp, elongate, tapering point, usually a leaf tip. Fig. 1, Fig. 472

Fig. 472

Adam-and-Eve, 69
Adam's-needle, 55
Adder's Tongue, 41
  Yellow, 40
Affinis Violet, 152
AGAVACEAE, 55
(*Agoseris cuspidata*), 214
Agrostemma
- ☐ *githago*, 77
AIZOACEAE, 75
Alexander
  Golden, 164
Alfalfa, 138
ALISMATACEAE, 30
Alleghany Mountain Spurge, 147
Alliaria
  (*officinalis*), 117
- ☐ *petiolata*, 117
Allium
- ☐ *canadense*, 41
- ☐ *cuthbertii*, 41
  (*tricoccum*), 41
Alsike Clover, 140
ALTERNATE: the leaf arrangement in which leaves arise singly at the nodes. Fig. 4, Fig. 473

Fig. 473

Alum Root
  Common, 124
AMARYLLIDACEAE, 54
AMARYLLIS FAMILY, 54
American Brooklime, 197
American Cowslip, 167
American Dog Violet, 156
American Globeflower, 101
American Jacob's-ladder, 177
American White Hellebore, 49
Amerorchis
- ☐ *rotundifolia*, 67
Amsonia, 172
Amsonia
- ☐ *tabernaemontana*, 172
Anagallis
- ☐ *arvensis*, 169
  var. *arvensis*, 169
  var. *coerulea*, 169
Androsace, 168
Androsace
- ☐ *occidentalis*, 168
Anemone
  Canada, 98
  Carolina, 98
  False Rue, 100
  Rue, 96
  Wood, 99
Anemone
- ☐ *canadensis*, 98

- ☐ *caroliniana*, 98
  (*patens*), 97
- ☐ *quinquefolia*, 99, 100
- ☐ *virginiana*, 99
  (*Anemonella thalictroides*), 96
ANGIOSPERMS, ANGIOSPERMAE, 6, 18
Anise Root, 162
Antelope-horn
  Green, 174
Antennaria
- ☐ *neglecta*, 217
- ☐ *plantaginifolia*, 216
Anthemis
- ☐ *arvensis*, 223
- ☐ *cotula*, 223
ANTHER: the pollen-containing part of the stamen. Fig. 7, Fig. 474

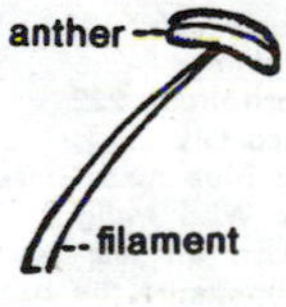

Fig. 474

APETALOUS: without petals.
(APIACEAE), 160
APICAL: at the tip (apex).
Aplectrum
- ☐ *hyemale*, 69
APOCYNACEAE, 172
Apocynum, 175
- ☐ *cannabinum*, 173
APPENDAGE: a projection from a larger structure. Fig. 383, Fig. 385
Appendaged Waterleaf, 182

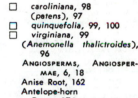

231

Aquilegia
☐ canadensis, 87
Arabis
☐ canadensis, 118
☐ glabra, 119
☐ hirsuta, 119
☐ laevigata, 119
☐ lyrata, 118
☐ shortii, 119
ARACEAE, 31
Aralia
☐ nudicaulis, 160
ARALIACEAE, 159
Arbutus
Trailing, 167
Arenaria
(lateriflora), 81
☐ serpyllifolia, 80
Arethusa, 66
Arethusa
☐ bulbosa, 66
Arisaema
☐ dracontium, 33
☐ triphyllum, 33
ARISTOLOCHIACEAE, 71
Arnica
☐ acaulis, 220
arrangement of collection, 13
Arrow-arum
Green, 32
Arrowhead
Common, 30
Arrow-leaf Sweet Coltsfoot, 224
Arrow-leaved Violet, 154
ARUM FAMILY, 31
Asarum
☐ canadense, 71
var. acuminatum, 71
var. reflexum, 71
ASCLEPIADACEAE, 173
Asclepias, 173
☐ amplexicaulis, 175
☐ quadrifolia, 174
☐ syriaca, 175
☐ tuberosa, 175
☐ verticillata, 174
☐ viridis, 174
Asparagus, 37
Asparagus
☐ officinalis, 37
(ASTERACEAE), 212
Astragalus
☐ crassicarpus, 135
☐ lotiflorus, 135
☐ tennesseensis, 135
Astranthium
☐ integrifolium, 222
Atamasco Lily, 54
Atlantic Blue-eyed Grass, 56
Atlantic Wild Indigo, 138

AURICLE: an ear-like appendage at the base of a leaf, sepal, or petal. Fig. 253
AURICULATE: with auricles.
author name, 7
Avens
Purple, 131
Spring, 130
White, 130
AXIL: the angle formed between two structures, particularly the leaf and the stem. Fig. 475

Fig. 475

AXILLARY: borne in the leaf axil.

B

Balsam Goundsel, 219
Baneberry
Red, 94
White, 95
Baptisia
☐ lactea, 138
(leucantha), 138
☐ leucophaea, 137
Barbarea
☐ vulgaris, 110
BARBERRY FAMILY, 102
Barren Strawberry, 127
BASAL: borne at the base, e.g. leaves at the base of a flowering stem (scape). Fig. 52-c, Fig. 476

Fig. 476

Bashful Trillium, 47
Bastard Toad-flax, 71
Beach Pea, 133
BEAK: the long tip on fruits such as an achene which is formed by a point on the ovary or the persistent style. Fig. 477

Fig. 477

BEARDED: bearing a tuft of hairs.
Beard-tongue
Cobaea, 202
Funnelform, 201
Hairy, 202
Large-flowered, 201
Beargrass, 55
Bedstraw, 210
Northern, 209
Southwestern, 209
Sweet-scented, 209
Yellow, 208
Beefsteak Plant, 196
Bellis
☐ perennis, 222
Bellwort
Large-flowered, 52
Mountain, 51
Perfoliate, 51
BERBERIDACEAE, 102
BERRY: a pulpy, indehiscent few- to many-seeded fruit.
Betony
Wood, 196
BIENNIAL: a plant which lives for two years and flowers during its second year.
BIFURCATE: deeply lobed into two parts; forked.
BILABIATE: two-lipped; with sepals or petals united so that the calyx or corolla is bilaterally symmetrical with an upper and a lower portion (as in the Mint family). Fig. 478

Fig. 478

BILATERALLY SYMMETRICAL: zygomorphic; capable of being divided into equal halves along only one plane. Fig. 20, Fig. 535
Bindweed, 176
Hedge, 176
binomial nomenclature, 7
Birdfoot Violet, 151
Bird's-eye, 198
Birthwort, 45
Birthwort Family, 71
Bishop's-cap, 124
Naked, 124
Bitter-cress
Hairy, 121
Meadow, 121
Pennsylvania, 116
Small-flowered, 121
Bitter Nightshade, 194

Bittersweet, 194
Black Bindweed, 72
Black Medic, 138
Black Mustard, 111
Black Snakeroot, 161
Bladder Campion, 78
BLADE: the broad portion of a leaf or petal. Fig. 2, Fig. 479

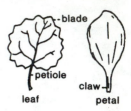

Fig. 479

Blazing-star, 37
Bleeding-heart, (105)
Wild, 105
Bloodroot, 103
Bluebell, 185
BLUEBELL FAMILY, 211
Bluebonnet
Texas, 136
Blue Cohosh, 102
Blue-eyed Grass, 56
Atlantic, 56
Pointed, 56
Blue-eyed Mary, 200
Blue Marsh Violet, 153
Blue-star, 172
Blue Toadflax, 195
Bluets, 206
Least, 207
Thyme-leaved, 206
Bluntleaf Milkweed, 175
Blunt-leaved Sandwort, 81
Bog Violet, 153
BORAGE FAMILY, 185
BORAGINACEAE, 185
BOX FAMILY, 147
BRACT: a modified leaf, often associated with a flower or inflorescence.
BRACTEATE: with bracts.
Bracted Spiderwort, 34
Brassica
(hirta), 111
☐ juncea, 111
(kaber), 111
(napus), 111
☐ nigra, 111
(oleracea), 111
(rapa), 111
(BRASSICACEAE), 108
Bread-root, 136
Broadleaf Dock, 74
BROMELIACEAE, 36
Brooklime
American, 197
Broom-rape
Naked, 202
BROOM-RAPE FAMILY, 202
(Brussels Sprouts), 111
Buckbean, 171
BUCKBEAN FAMILY, 171
Buckwheat, (72)
Wild, 72

BUCKWHEAT FAMILY, 72
Buglossoides
□ *arvense*, 186
BULB: a very short under-
    ground stem bearing
    fleshy modified leaves
    as in an onion.
BULBLET: a small bulb, usu-
    ally borne on a stem.
Bunchberry, 166
Butter-and-eggs, 195
Butterbur, 224
Buttercup
    Creeping, 91
    Dwarf, 93
    Early, 92
    Hispid, 93
    Swamp, 92
    Tall, 91
BUTTERCUP FAMILY, 85
Butterfly Dock, 224
Butterflyweed, 175
BUXACEAE, 147

C

Cabbage, (111)
    Skunk, 31
CABBAGE FAMILY, 108
Calla
    Wild, 32
Calla
□ *palustris*, 32
Calochortus
□ *nuttallii*, 49
Caltha
□ *palustris*, 88
Calylophus
□ *serrulatus*, 158
Calypso, 65
Calypso
□ *bulbosa*, 65
Calystegia
□ *sepium*, 176
CALYX: a collective term for
    all of the sepals of a
    flower. Fig. 5, Fig. 480

Fig. 480

Camassia
    (*esculenta*), 43
□ *scilloides*, 43
Camelina
□ *microcarpa*, 108
□ *sativa*, 109
Campanula
□ *rotundifolia*, 211
CAMPANULACEAE, 211
CAMPANULATE: bell-
    shaped.
Campion
    Bladder, 78
    White, 79

Canada Anemone, 98
Canada Mayflower, 38
Canada Violet, 157
Canada Waterleaf, 183
Cancer-root, 203
    One-flowered, 202
CAPITATE: bearing a head-
    like knob on a stalk, or
    with a series of flowers
    collected into a head or
    dense cluster.
CAPRIFOLIACEAE, 210
Capsella
□ *bursa-pastoris*, 114
CAPSULE: a dry, dehiscent
    fruit of more than one
    carpel. Fig. 11
Cardamine
□ *bulbosa*, 120
□ *douglassii*, 120
□ *hirsuta*, 121
□ *parviflora*, 121
□ *pensylvanica*, 116
□ *pratensis*, 121
(Carnation), 77
carnivorous plants, 122
Carolina Anemone, 98
Carolina Dwarf Dandelion,
    215
Carolina Larkspur, 87
Carolina Pink, 170
Carolina Spring Beauty, 76
Carolina Vetch, 132
Carolina Whitlow-grass, 112
CARPEL: a simple pistil or
    one part of a compound
    pistil. Fig. 8, Fig. 481

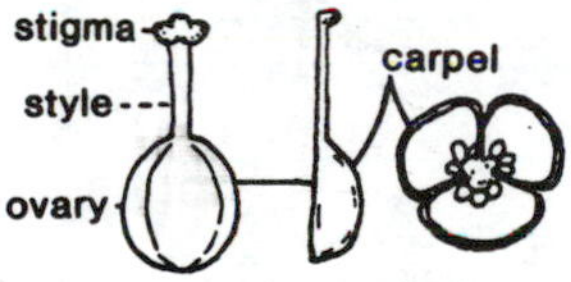

stigma
style
ovary
carpel

Fig. 481

Carpetweed, 75
CARPET WEED FAMILY, 75
CARYOPHYLLACEAE, 77
Castilleja
□ *coccinea*, 195
□ *sessiliflora*, 195
(Castor-bean), 145
Catchfly
    Night-flowering, 79
    Sleepy, 80
CAULINE: borne on the
    stem.
Caulophyllum
□ *thalictroides*, 102
Celandine, 104
Celandine Poppy, 104
CENTURY PLANT FAMILY, 55
Cerastium, 81
□ *arvense*, 82
□ *fontanum* ssp. *triviale*, 84
□ *glomeratum*, 83
□ *nutans*, 83
    var. *brachypodum*, 83
    (*viscosum*), 83
    (*vulgatum*), 84
Chaerophyllum
□ *procumbens*, 162

Chamaelirium
□ *luteum*, 37
(*Chamaesyce*), 145
Charlock, 111
Cheeses, 148
Chelidonium
□ *majus*, 104
Chervil
    Spreading, 162
Chickweed
    Common, 81
    Common Mouse-ear, 84
    Field, 82
    Long-leaved, 82
    Nodding, 83
    Short-stalked, 83
    Sticky Mouse-ear, 83
    Water Mouse-ear, 82
Chickweed Phlox, 180
Chicory, 214
(*Chrysanthemum leucanthe-*
    mum), 223
Chrysogonum, 220
Chrysogonum
□ *virginianum*, 220
    var. *australe*, 220
Chrysosplenium
□ *americanum*, 123
Cichorium
□ *intybus*, 214
(Cineraria), 219
Cinquefoil
    Common, 128
    Rough-fruited, 129
    Silvery, 129
CISTACEAE, 148
Clammy Ground Cherry, 194
CLASPING: wholly or part-
    ly encircling the stem.
Clasping-leaved Twisted-
    stalk, 53
CLAW: the narrowed basal
    part of some petals or
    sepals. Fig. 479
Claytonia
□ *caroliniana*, 76
□ *lanceolata*, 76
□ *virginica*, 76
Cleavers, 210
Cleft Phlox, 180
CLEISTOGAMOUS: descrip-
    tive of a flower that
    does not open and is
    self-fertilized.
Clintonia
    White, 42
    Yellow, 42
Clintonia
□ *borealis*, 42
□ *umbellulata*, 42
□ *uniflora*, 42
Clover
    Alsike, 140
    Crimson, 139
    Low Hop, 139
    Rabbit-foot, 139
    Red, 140
    White, 140
    White Sweet, 137
    Yellow, 139
Cobaea Beard-tongue, 202
Cockle
    Corn, 77
Coeloglossum
□ *viride* var. *virescens*, 66
Cohosh
    Blue, 102
collecting, 10-11

Collinsia
    Narrow-leaved, 200
Collinsia
□ *verna*, 200
□ *violacea*, 200
Coltsfoot, 217
    Arrow-leaf Sweet, 224
    Sweet, 216, 224
Columbine, 87
    Wild, 87
COLUMN: the central struc-
    ture in orchid flowers
    consisting of fused fila-
    ments and style. Fig.
    136
Comandra
    Pale, 71
Comandra
□ *umbellata*, 71
    ssp. *pallida*, 71
Comfrey
    Wild, 185
Commelina
□ *erecta*, 34
COMMELINACEAE, 34
Common Alum Root, 124
Common Arrowhead, 30
Common Blue Violet, 153
Common Chickweed, 81
Common Cinquefoil, 128
Common Groundsel, 216
Common Lousewort, 196
Common Mallow, 148
Common Milkweed, 175
Common Mouse-ear Chick-
    weed, 84
Common Plantain, 205
Common Speedwell, 197
COMPOSITAE, 212
Composite Family, 212
COMPOUND LEAF: a leaf
    divided into several
    leaflets. Fig. 3, Fig. 482

Fig. 482

Confederate Violet, 153
(*Conium maculatum*), 162
CONNATE: with similar
    structures grown togeth-
    er.
Conopholis
□ *americana*, 203
Conringia
□ *orientalis*, 109
Convallaria
□ *majalis*, 40
CONVOLVULACEAE, 176
Convolvulus
□ *arvensis*, 176
    (*sepium*), 176
Copper Iris, 57
Coptis
□ *trifolia*, 101
(Coralbells), 124
Corallorhiza
□ *odontorhiza*, 63
□ *striata*, 64

□ *trifida*, 63
□ *wisteriana*, 63
Coral-root
  Early, 63
  Late Southern, 63
  Spring, 63
  Striped, 64
  Wister's, 63
CORDATE: heart-shaped. Fig. 483

Fig. 483

Coreopsis
□ *lanceolata*, 221
CORM: a short, swollen underground stem, sometimes covered with thin scaly leaves.
Cornaceae, 166
Corn Cockle, 77
Corn Gromwell, 186
Corn Lily, 42
Corn Speedwell, 199
Cornus
□ *canadensis*, 166
  (*florida*), 166
COROLLA: the collective term for all of the petals of a flower. Fig. 5, Fig. 484

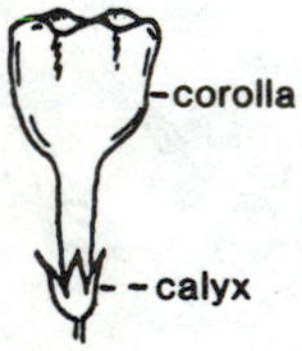

—corolla

—calyx

Fig. 484

CORONA: an appendage between the stamens and the perianth, as in the milkweeds. Fig. 363, Fig. 485

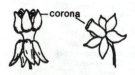

—corona

Fig. 485

Corydalis
  Golden, 107
  Pale, 106
  Vesicular, 108
Corydalis, 104

□ *aurea*, 107
□ *crystallina*, 108
□ *micrantha*, 107
□ *sempervirens*, 106
CORYMB: a flat-topped flower cluster blooming from the outside to the inside. Fig. 10, Fig. 486

Fig. 486

COTYLEDON: a first leaf of the embryo, present within the seed and sometimes modified for food storage.
Cow Cress, 115
Cow Parsnip, 161
Cowslip
  American, 167
  Virginia, 185
Cow Vetch, 133
Cranesbill, 144
Crassulaceae, 123
CREEPING: a plant growth pattern which spreads over the surface of the ground and produces roots at the nodes.
Creeping Buttercup, 91
Creeping Phlox, 179
Creeping Snowberry, 166
Creeping Wood Sorrel, 142
CRENATE: a margin with rounded teeth. Fig. 1, Fig. 487

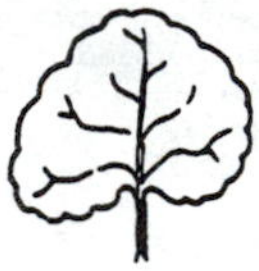

Fig. 487

Cress
  Bitter-
    Hairy, 121
    Meadow, 121
    Pennsylvania, 116
    Small-flowered, 121
  Cow, 115
  Field, 115
  Penny-
    Field, 114
    Perfoliate, 120
  Purple, 120
  Rock-
    Hairy, 119
    Lyre-leaved, 118
    Smooth, 119

  Toothed, 119
  Spring, 120
  Watercress, 116
  Yellow, 110
CREST: a short ridge-like projection from the surface. Fig. 231
Crimson Clover, 139
Crinkleroot, 113
(Crocus), 59
Crowfoot
  Cursed, 89
  Hooked, 90
  Small-flowered, 94
  Yellow Water-, 89
Cruciferae, 108
Cucumber-root
  Indian, 44
Curled Dock, 74
Cursed Crowfoot, 89
Cut-leaved Toothwort, 113
(Cyclamen), 167
CYME: a somewhat flat-topped flower cluster that blooms from the inside to the outside, with the terminal flower maturing first. Fig. 10, Fig. 488

Fig. 488

Cynoglossum
□ *officinale*, 186
□ *virginianum*, 185
Cypress Spurge, 146
Cypripedium
□ *acaule*, 60
□ *arietinum*, 61
  (*calceolus var. pubescens*), 62
□ *candidum*, 62
□ *pubescens*, 62
□ *reginae*, 61

D

(Daffodil), 54
Daisy
  English, 222
  Lawn, 222
  Ox-eye, 223
  Western, 222
Daisy Fleabane, 225
Dame's-rocket, 116
Dame's Violet, 116
Dandelion, 213
  Dwarf, 215
    Carolina, 215
    Western, 214
  False
    Leafy-stemmed, 213
    Prairie, 214
    Rough, 213

  Potato, 215
  Red Seeded, 213
Dayflower, 34
Dead Nattle, 193
Death Camas, 48
DECIDUOUS: falling, as the leaves of broad-leaved trees; not evergreen.
DECUMBENT: with stems lying on the ground but the tips ascending. Fig. 489

Fig. 489

DEHISCENCE: the manner or process of opening of a fruit or anther.
DEHISCENT: opening when mature.
Delphinium
□ *carolinianum*, 87
□ *tricorne*, 86
□ *virescens*, 86
Dentaria
□ *diphylla*, 113
□ *heterophylla*, 113
□ *laciniata*, 113
□ *maxima*, 113
DENTATE: with pointed teeth directed outward. Fig. 1
Devil's-bit, 37
Dicentra, 104
□ *canadensis*, 106
□ *cucullaria*, 105
□ *eximia*, 105
□ *spectabilis*, 105
Dicotyledons, Dicotyledonae, 6, 71
(*Dieffenbachia*), 33
DIOECIOUS: with staminate (male) and pistillate (female) flowers on separate plants.
(*Dionaea*), 122
DISK FLOWER: the radially symmetrical, tubular flower of the Compositae. Fig. 443, Fig. 445, Fig. 490

Fig. 490

Disporum
  Hairy, 52
  Rough-fruited, 52
Disporum
□ *lanuginosum*, 52
□ *trachycarpum*, 52

DISSECTED: many times divided into narrow segments.
Dock
  Broadleaf, 74
  Butterfly, 224
  Curled, 74
  (Patience), 74
  Peach-leaved, 74
Dodecatheon
  meadia, 167
  (Dogbane), 175
DOGBANE FAMILY, 172
Dog Fennel, 223
Dog-tooth Violet, 41
DOGWOOD FAMILY, 166
Doll's-eyes, 95
Douglas' Phlox, 179
Downy Painted-cup, 195
Downy Phlox, 181
Downy Yellow Violet, 155
Draba
  brachycarpa, 112
  cuneifolia, 112
  reptans, 112
  (verna), 112
Dragon-mouth, 66
Drooping Star-of-Bethlehem, 43
(Drosera), 122
DRUPE: a fleshy fruit with a pit, as in the peach.
drying specimens, 12-13
Dry Strawberry, 127
Duchesnea
  indica, 127
  (Dusty-miller), 219
Dutchman's-breeches, 105
Dwarf Buttercup, 93
Dwarf Crested Iris, 57
Dwarf Dandelion, 215
Dwarf European Iris, 57
Dwarf Ginseng, 159
Dwarf Iris, 58
Dwarf Larkspur, 86
Dwarf White Trillium, 47

E

Early Blue Violet, 154
Early Buttercup, 92
Early Coral-root, 63
Early Meadow Parsnip, 165
Early Meadow Rue, 95
Early Saxifrage, 126
edible plants, 8-9
Eichornia
  crassipes, 36
Ellisia
  nyctelea, 182
EMARGINATE: bearing a broad, shallow notch at the tip. Fig. 1
endangered plants, 14-16
English Daisy, 222
English Plantain, 204
ENTIRE: a margin without divisions or teeth. Fig. 491

Fig. 491

EPHEMERAL: lasting for only a short time, e.g. a day or less.
Epigaea
  repens, 167
EPIPHYTE: a plant which grows on another plant for support but obtains its nutrients and water from the air.
ERICACEAE, 166
Erigenia
  bulbosa, 163
Erigeron
  philadelphicus, 225
  pulchellus, 224
Erodium
  cicutarium, 142
Erophila
  verna, 112
Erysimum
  asperum, 109
  cheiranthoides, 109
Erythronium
  albidum, 41
  americanum, 40
Euphorbia
  commutata, 146
  corollata, 145
  cyparissias, 146
EUPHORBIACEAE, 145
European Field Pansy, 155
Evening Primrose
  Three-lobe, 159
  Toothleaf, 158
  White, 159
EVENING PRIMROSE FAMILY, 158
Everlasting, 217
EXSERTED: protruding beyond the calyx or corolla. Fig. 492

Fig. 492

F

(FABACEAE), 132
(*Fagopyrum esculentum*), 72
Fairy-slipper, 65
False Flax, 108
False Garlic, 42
False Mermaid, 147
FALSE MERMAID FAMILY, 147
False Mitrewort, 125

False Rue Anemone, 100
False Solomon's-seal, 50
Fawn Lily
  White, 41
Fennel
  Dog, 223
FIBROUS ROOTS: roots which are thin and branching, with no single main root.
Field Chickweed, 82
Field Cress, 115
Field Penny-cress, 114
Figwort
  Hare, 200
  Maryland, 200
FIGWORT FAMILY, 195
FILAMENT: the structure that supports the anther. Fig. 7, Fig. 493

anther-
-filament

Fig. 493

Filaree, 142
FILIFORM: thread-like; slender and cylindrical.
Fire Pink, 77
first aid for poisoning, 10
Flag
  Large Blue, 59
  Slender Blue, 58
  Southern Blue, 59
  Sweet, 31
Flax
  False, 108
Fleabane
  Daisy, 225
  Philadelphia, 225
Floerkea
  proserpinacoides, 147
(Flowering Dogwood), 166
Flowering Spurge, 145
Flowering Wintergreen, 144
flower parts, 3-4
Foamflower, 125
FOLIOLATE: with leaflets; often used with a number to indicate the number of leaflets in a compound leaf, as 6-foliolate. Fig. 494

Fig. 494

FOLLICLE: a dry fruit composed of a single carpel and opening along one side. Fig. 11, Fig. 495

Fig. 495

Forget-me-not, 187
  Smaller, 187
Four-leaved Milkweed, 174
Four-o'clock
  Wild, 75
FOUR O'CLOCK FAMILY, 75
Fragaria
  vesca, 126
  virginiana, 126
(*Fragaria chiloensis*), 126
Fragrant Waterlily, 85
Fringed Milkwort, 144
Fringed Phacelia, 184
Frostweed, 148
FRUIT: the mature ovary, sometimes with additional floral parts, and its seeds.
FUMARIACEAE, 104
Fumeroot
  Golden, 107
Fumewort
  Slender, 107
FUMITORY FAMILY, 104
FUNNELFORM: funnel shaped; with a wide diameter at the mouth and tapering to a smaller diameter at the base.
Funnelform Beard-tongue, 201
FUSIFORM: football shaped; of larger diameter in the middle and tapering to both ends.

G

Galearis
  spectabilis, 67
Galinsoga, 221
Galinsoga
  (ciliata), 221
  parviflora, 221
  quadriradiata, 221
Galium
  aparine, 210
  boreale, 209
  circaezans, 209
  obtusum, 209
  tinctorium, 210
  triflorum, 209
  verum, 208
  virgatum, 209
Garden Verbena, 189
Garlic,
  False, 42
  Meadow, 41
  Striped, 41

Garlic Mustard, 117
Gaultheria
☐ *hispidula*, 166
☐ *procumbens*, 166
Gay-wings, 144
(Gentian), 171
GENTIANACEAE, 171
GENTIAN FAMILY, 171
genus, genera, 6, 7
genus cover, 13
GERANIACEAE, 142
Geranium
Wild, 143
Geranium
☐ *carolinianum*, 144
☐ *maculatum*, 143
☐ *robertianum*, 143
GERANIUM FAMILY, 142
Geum, 127
☐ *canadense*, 130
☐ *rivale*, 131
☐ *triflorum*, 131
☐ *vernum*, 130
Gill-over-the-ground, 192
Ginger
Wild, 71
Ginseng
Dwarf, 159
GINSENG FAMILY, 159
GLABROUS: not hairy.
(Gladiolus), 59
GLAND: a secreting struc-
ture, as a nectar gland
or the tip of a hair.
GLANDULAR: bearing
glands.
GLAUCOUS: covered with a
whitish film.
Glechoma
☐ *hederacea*, 192
Globeflower
American, 101
GLOBOSE: round, like a
globe.
Goat's Rue, 134
Goldcups, 90
Golden Alexander, 164
Golden Corydalis, 107
Golden Fumeroot, 107
Golden Parsnip, 163
Golden Ragwort, 218
Golden Saxifrage, 123
Golden-seal, 100
Golden-slipper, 62
Goldthread, 101
Goose Grass, 210
Ground Cherry
Clammy, 194
Virginia, 194
Ground Ivy, 192
Ground Pink, 177
Ground Plum, 135
Groundsel
Balsam, 219
Common, 216
Grape Hyacinth, 39
Grass
Blue-eyed, 56
Atlantic, 56
Pointed, 56
Goose, 210
Whitlow-
Carolina, 112
Short-fruited, 112
Vernal, 112
Wedge-leaved, 112
Worm-, 170
Yellow Star-, 54

Gratiola
☐ *neglecta*, 196
Great Solomon's-seal, 38
Greek Valerian, 177
Green-and-gold, 220
Green Antelope-horn, 174
Green Arrow-arum, 32
Green-dragon, 33
Green Violet, 149
Gromwell
Corn, 186
GYNOSTEGIUM: the fused
anthers and stigma of
Milkweed flowers. Fig.
363

H

(*Habenaria viridis* var. *brac-
teata*), 66
Hairy Beard-tongue, 202
Hairy Bitter-cress, 121
Hairy Disporum, 52
Hairy-jointed Meadow Par-
snip, 165
Hairy Phlox, 181
Hairy Puccoon, 188
Hairy Rock-cress, 119
Hairy Solomon's-seal, 39
Hairy Vetch, 132
Harbinger-of-Spring, 163
Harebell, 211
Hare Figwort, 200
Hare's-ear, 109
HASTATE: arrow-shaped but
with the basal lobes
pointing outward. Fig. 1
HEAD: an inflorescence in
which sessile flowers
are clustered tightly to-
gether on a short or
flattened axis. Fig. 10,
Fig. 443
Heal-all, 191
Heart-leaved Plantain, 205
Heart-leaved Twayblade, 69
HEATH FAMILY, 166
Hedeoma
☐ *hispidum*, 190
Hedge Bindweed, 176
Hedge Hyssop, 196
Hedge Mustard, 110
Hedyotis
Large, 207
Slender-leaved, 208
Hedyotis
☐ *caerulea*, 206
☐ *crassifolia*, 207
☐ *longifolia*, 208
☐ *nuttalliana*, 208
☐ *michauxii*, 206
☐ *purpurea*, 207
Helianthemum
☐ *canadense*, 148
Hellebore
American White, 49
Hemp
Indian, 173
Henbit, 193
Hepatica, 97
Hepatica
(*americana*), 97
☐ *nobilis*, 97
var. *obtusa*, 97
Heracleum
☐ *lanatum*, 161

HERBACEOUS: like an herb;
not woody.
herbarium, 10
herbarium specimen, 10-11
Herb Robert, 143
Heronbill, 142
Hesperis
☐ *matronalis*, 116
Heuchera
☐ *americana*, 124
var. *hispida*, 124
(*sanguinea*), 124
HIRSUTE: with stiff hairs.
HISPID: with very stiff
hairs or bristles.
Hispid Buttercup, 93
Hoary Puccoon, 188
(Hollyhock), 148
HONEYSUCKLE FAMILY, 210
HOOD: a structure which
overarches another struc-
ture, esp. one of the
parts of the corona in
Milkweed flowers. Fig.
363
Hood's Phlox, 179
Hooked Crowfoot, 90
HORN: a horn-like projec-
tion, esp. in Milkweed
flowers. Fig. 364
Horse Gentian, 210
Scarlet-fruited, 210
Horse Nettle, 194
Hound's-tongue, 186
(*Houstonia*), 206
(*caerulea*), 206
(*minima*), 207
(*patens*), 207
(*purpurea*), 207
(*serpyllifolia*), 206
(*tenuifolia*), 208
Hyacinth
Grape, 39
Starch Grape, 39
Water, 36
Wild, 43
Hybanthus
☐ *concolor*, 149
Hydrastis
☐ *canadensis*, 100
HYDROPHYLLACEAE, 181
Hydrophyllum
☐ *appendiculatum*, 182
☐ *canadense*, 183
☐ *macrophyllum*, 183
☐ *virginianum*, 183
HYPANTHIUM: the fleshy
cup-like structure in a
flower from the rim of
which the sepals, petals,
and stamens arise. Fig.
27.
Hypoxis
☐ *hirsuta*, 54
Hyssop
Hedge, 196

I

Ill-scented Wake-robin, 45
IMPERFECT FLOWER: a uni-
sexual flower, with
either functional sta-
mens or pistils.
INCLUDED: not protruding
beyond the calyx or
corolla. Fig. 496

Fig. 496

INDEHISCENT: not opening
at maturity.
Indian Cucumber-root, 44
Indian Hemp, 173
Indian Paintbrush, 195
Indian Pink, 77, 170
Indian Strawberry, 127
Indian Turnip, 33
Indian Wheat
Wolly, 204
Indigo
Atlantic Wild, 138
Large-bracted Wild, 137
INFERIOR OVARY: an ovary
that is positioned below
the apparent point of
attachment of the sepals
and petals. Fig. 497

Fig. 497

INFLORESCENCE: the collective name for a cluster
of flowers on a plant
stalk; clusters of flowers
arranged in particular
ways are specific types
of inflorescences. Fig.
10
Innocence, 206
INTERNODE: the part of the
stem between the nodes.
INVOLUCRE: a group of
bracts surrounding the
flower or cluster of
flowers (as in the Com-
posite family).
Iodanthus
☐ *pinnatifidus*, 117
IRIDACEAE, 55
Iris
Copper, 57
Dwarf, 58
Dwarf Crested, 57
Dwarf European, 57
Prairie, 55
Iris
☐ *cristata*, 57
☐ *fulva*, 57
(X *germanica*), 58
☐ *prismatica*, 58
☐ *pumila*, 57

  verna, 58
  versicolor, 59
  virginica, 59
IRIS FAMILY, 55
IRREGULAR: with similar parts unequal in size, shape or union; see zygomorphic. Fig. 498

Fig. 498

Isopyrum
  biternatum, 100
Isotria
  verticillata, 64
Ivy
  Ground, 192

J

Jack-in-the-pulpit, 33
Jacob's Ladder
  American, 177
Jeffersonia
  diphylla, 102
Johnny-jump-up, 155

K

KEEL: a longitudinal ridge like the keel of a boat, often formed from the fold of a petal or the union of two petals as in the Legume family.
Knotweed, 73
Krigia, 212
  biflora, 215
  dandelion, 215
  occidentalis, 214
  virginica, 215

L

label, 11, Fig. 13
LABIATAE, 190
Ladies' Sorrel, 141
Ladies'-tobacco, 217
Ladies'-tresses
  Slender, 68
  Spring, 68
Lady's-slipper
  Pink, 60
  Ram's Head, 61
  Showy, 61
  Stemless, 60
  White, 62
  Yellow, 62
Lady's-thumb, 72
(LAMIACEAE), 190
Lamium

  amplexicaule, 193
  purpureum, 193
Lance-leaved Tickseed, 221
Lance-leaved Violet, 150
LANCEOLATE: lance-shaped; much longer than broad, tapered at both ends and broadest below the middle. Fig. 499

Fig. 499

Large Blue Flag, 59
Large-bracted Plantain, 203
Large-bracted Wild Indigo, 137
Large-flowered Beard-tongue, 201
Large-flowered Bellwort, 52
Large-flowering Trillium, 46
Large Hedyotis, 207
(Large Houstonia), 207
Large-leaved Waterleaf, 183
Large Toothwort, 113
Larkspur
  Carolina, 87
  Dwarf, 86
  Prairie, 86
Late Southern Coral-root, 63
Lathyrus
  japonicus, 133
  venosus, 134
Lawn Daisy, 222
LEAFLET: one of the leaf-like divisions of a compound leaf. Fig. 3
Leaf Mustard, 111
leaf shapes, Fig. 1
Leafy-stemmed False Dandelion, 213
Least Bluets, 207
Leek
  Wild, 41
LEGUME: the bean or pea-like fruits (pods) of the Legume family, formed from a single carpel and opening along both sides. Fig. 11
LEGUME FAMILY, 132
LEGUMINOSAE, 132
Lemon
  Wild, 103
Leonurus
  cardiaca, 192
Leopard's-bane, 220
Lepidium
  campestre, 115
  virginicum, 115
Leucanthemum
  vulgare, 223
LILIACEAE, 37

Lily
  Atamasco, 54
  Corn, 42
  Fawn, 41
  Rain, 54
  Sego, 49
  Trout, 41
  Zephyr, 54
LILY FAMILY, 37
Lily-leaved Twayblade, 70
Lily-of-the-valley, 40
  Wild, 38
LIMB: the apical, usually spreading, portion of a tubular corolla or calyx. Fig. 500

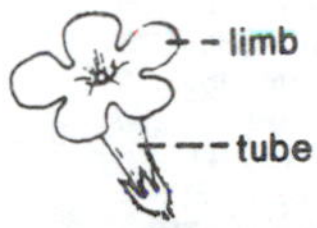

limb
tube

Fig. 500

LIMNANTHACEAE, 147
Linaria
  canadensis, 195
  vulgaris, 195
LINEAR: a long, narrow shape in which the margins are parallel or nearly so. Fig. 1
LIP: a modified petal in the Orchid family (Fig. 136) or the united upper or lower petals or sepals of an irregular flower such as in the Mint family. Fig. 501

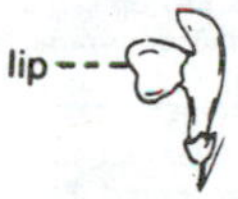

lip

Fig. 501

Liparis
  lilifolia, 70
  loeselii, 70
Liquorice
  Wild, 209
Listera
  cordata, 69
Lithospermum
  (arvense), 186
  canescens, 188
  caroliniense, 188
  (croceum), 188
  incisum, 187
Liverwort, 97
LOBED: with large partial divisions, as in the margins of a maple or oak leaf. Fig. 1
LOCULE: the space inside a fruit or ovary in which the seeds develop, Fig. 502; in an anther, the space in which the pollen develops.

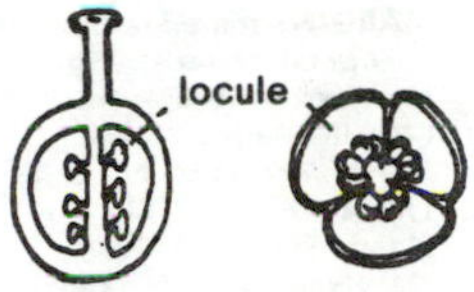

locule

Fig. 502

Loesel's Twayblade, 70
LOGANIACEAE, 170
LOGANIA FAMILY, 170
Long-bracted Orchis, 66
Long-leaved Chickweed, 82
Long-spurred Violet, 156
Loose-flowered Phacelia, 184
Loosestrife
  Tufted, 170
Lotus, 85
Lousewort
  Common, 196
Low Hop Clover, 139
Low Milk-vetch, 135
Lupine
  Wild, 136
Lupinus
  perennis, 136
  (subcarinosus), 136
  (texensis), 136
(Lychnis alba), 79
LYRATE: elongate with a large round terminal lobe and smaller lower ones; shaped like a lyre. Fig. 503

Fig. 503

Lyre-leaved Rock-cress, 118
Lyre-leaved Sage, 191
Lysimachia
  nummularia, 169
  quadrifolia, 169
  thyrsiflora, 170

M

MADDER FAMILY, 205
Maianthemum
  canadense, 38
Mallow
  Common, 148
  (Marsh), 148
MALLOW FAMILY, 148
Malva
  neglecta, 148
MALVACEAE, 148
Mandarin
  Yellow, 52
Mandrake
  Wild, 103

MARGIN: the edge of a leaf, petal or sepal. Fig. 1
Marguerite, 223
(Marsh Mallow), 148
Marsh Marigold, 88
(Marsh Pink), 171
Marsh Trefoil, 171
Maryland Figwort, 200
May Apple, 103
Mayflower, 167
   Canada, 38
Maypop, 158
Meadow Bitter-cress, 121
Meadow Garlic, 41
Meadow-gold, 90
Meadow Rue
   Early, 95
   Purple, 96
   Waxy, 96
Medeola
☐ *virginiana*, 44
Medic
   Black, 138
Medicago
☐ *lupulina*, 138
☐ *sativa*, 138
medicinal plants, 8
Melilotus, 137
Melilotus
☐ *alba*, 137
☐ *officinalis*, 137
MEMBRANOUS: thin and translucent like a membrane.
MENYANTHACEAE, 171
Menyanthes
☐ *trifoliata*, 171
Mermaid
   False, 147
Merry-hearts, 48
Mertensia
☐ *virginica*, 185
MIDRIB: the main vein of a leaf.
Milfoil, 222
Milk-vetch
   Low, 135
   Tennessee, 135
Milkweed
   Bluntleaf, 175
   Common, 175
   Four-leaved, 174
   Whorled, 174
MILKWEED FAMILY, 173
Milkwort
   Fringed, 144
   Racemed, 145
MILKWORT FAMILY, 144
MINT FAMILY, 190
Mirabilis
☐ *nyctaginea*, 75
Missouri Violet, 152
Mitchella
☐ *repens*, 206
Mitella
☐ *diphylla*, 124
☐ *nuda*, 124
Mitrewort, 124
   False, 125
Moccasin Flower, 60
Mock Pennyroyal, 190
Moehringia
☐ *lateriflora*, 81
Mollugo
☐ *verticillata*, 75
Moneywort, 169
MONOCOTYLEDONS, MONOCO-TYLEDONAE, 6, 30

MONOECIOUS: with stamens and pistils in separate flowers on the same plant.
MORNING-GLORY FAMILY, 176
Moss Phlox, 179
Moss Pink, 177
Motherwort, 192
MOTTLED: with spots or blotches of a different color.
Mountain Bellwort, 51
Mountain Phlox, 178
mounting, 13
Mousetail, 88
Moxie Plum, 166
MUCRONATE: with a short, abrupt tip. Fig. 1
Muscari
☐ *atlanticum*, 39
☐ *botryoides*, 39
Mustard
   Black, 111
   Garlic, 117
   Hedge, 110
   Leaf, 111
   Tower, 119
   Treacle, 109
   White, 111
   Wormseed, 109
(MUSTARD FAMILY), 108
Myosotis
☐ *laxa*, 187
☐ *scorpioides*, 187
☐ *verna*, 187
Myosoton
☐ *aquaticum*, 82
Myosurus
☐ *minimus*, 88
Myrtle, 172

N

Naked Bishop's-cap, 124
Naked Broom-rape, 202
names, 7
(Narcissus), 54
Narrow-leaved Collinsia, 200
Narrow-leaved Puccoon, 187
Narrow-leaved Sagittaria, 30
Nasturtium
☐ *officinale*, 116
Native Watercress, 116
Neckweed, 199
NECTARY: the structure in which nectar is secreted.
Nelumbo
☐ *lutea*, 85
Nemastylis
☐ *geminiflora*, 55
Nemophila
   Small-flowered, 181
Nemophila
☐ *aphylla*, 181
   (*microcalyx*), 181
Nettle
   Dead, 193
   Horse, 194
NET-VEINED: with veins crossing and fusing with one another like strands of a net. Fig. 2
Night-flowering Catchfly, 79
Nightshade
   Bitter, 194
NIGHTSHADE FAMILY, 194

NODDING: with the flowers drooping so that the face of the flower points downwards.
Nodding Chickweed, 83
Nodding Trillium, 46
NODE: the point on a stem at which a leaf is attached. Fig. 4, 504

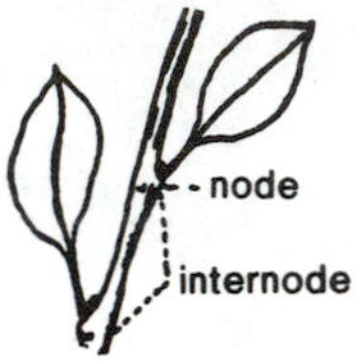

node
internode

Fig. 504

nomenclature, 7
Northern Bedstraw, 209
Nothocalais
☐ *cuspidata*, 214
Nothoscordum
☐ *bivalve*, 42
Nuphar
   (*advena*), 84
☐ *luteum*, 84
NUTLET: a small 1-seeded indehiscent fruit with a thick wall, as in the Mint family. Fig. 11
NYCTAGINACEAE, 75
Nyctelea, 182
Nymphaea
☐ *odorata*, 85
NYMPHAEACEAE, 84

O

Oats
   Wild, 51
OB-: a prefix used to indicate a condition reversed from the usual, as a shape in which the larger end is toward the tip, e.g. obovate. Fig. 505

Fig. 505

Obolaria
☐ *virginica*, 171
OBTUSE: rounded, not pointed.
Oenothera

(*serrulata*), 158
☐ *speciosa*, 159
☐ *triloba*, 159
Ohio Spiderwort, 35
(Okra), 148
ONAGRACEAE, 158
One-flowered Cancer-root, 202
Onion
   Wild, 48
OPPOSITE: the leaf arrangement in which 2 leaves arise from the same node, on opposite sides of the stem. Fig. 4
ORBICULAR: circular.
ORCHID FAMILY, 60
ORCHIDACEAE, 60
Orchis
   Long-bracted, 66
   Round-leaved, 67
   Showy, 67
(*Orchis*)
   (*rotundifolia*), 67
   (*spectabilis*), 67
Ornithagalum
☐ *nutans*, 43
☐ *umbellatum*, 43
OROBANCHACEAE, 202
Orobanche
☐ *uniflora*, 202
Osmorhiza
☐ *claytonii*, 162
☐ *longistylis*, 162
OVARY: the lower part of the pistil that contains the ovules and later the seeds. Fig. 7, Fig. 506

ovary

Fig. 506

OVATE: egg-shaped in outline, with the larger end toward the base, Fig. 507. Obovate: egg-shaped with the broader end toward the tip.

Fig. 507

OVOID: oval-shaped and three-dimensional.
OVULE: one of the bodies

in the ovary which contains the egg cell and becomes the seed. 5, Fig. 7, Fig. 508

Fig. 508

OXALIDACEAE, 141
Oxalis
☐ *corniculata*, 142
☐ *stricta*, 141
☐ *violacea*, 141
Ox-eye Daisy, 223
(*Oxybaphus*), 75

P

Pachysandra, 147
Pachysandra
☐ *procumbens*, 147
☐ *terminalis*, 147
Painted-cup
  Downy, 195
  Scarlet, 195
Painted Trillium, 48
Pale Comandra, 71
Pale Corydalis, 106
Pale Violet, 157
PALMATE: lobed, divided, or veined in a finger-like fashion with the divisions or veins arising from a common point. Fig. 3, Fig. 509

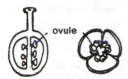

Fig. 509

Panax
  (ginseng), 159
☐ *quinquefolius*, 159
☐ *trifolius*, 159
PANICLE: an inflorescence in which the lateral branches rebranch forming a loose flower cluster longer than wide with the older flowers lower. Fig. 10, Fig. 510

Fig. 510

Pansy
  European Field, 155
  Wild, 155
pansy hill, 151
PAPAVERACEAE, 103
Papoose-root, 102
PAPPUS: the modified calyx of the Compositae, borne at the summit of the inferior ovary, consisting of hairs, scales or bristles. Fig. 276, Fig. 511

Fig. 511

PARALLEL VEINED: with the main veins running parallel to each other from base to apex. Fig. 2
PARSLEY FAMILY, 160
Parsnip
  Cow, 161
  Golden, 163
  Meadow
    Early, 165
    Hairy-jointed, 165
PARTED: deeply divided into lobes or segments.
Partridge-berry, 206
Pasque Flower, 97
Passiflora
☐ *incarnata*, 158
PASSIFLORACEAE, 158
Passion Flower, 158
PASSION FLOWER FAMILY, 158
(Patience Dock), 74
Pea
  Beach, 133
  Sweetpea, Wild, 134
  Veiny, 134
Peach-leaved Dock, 74
PECTINATE: comb-like, divided into narrow parallel segments like the teeth of a comb. Fig. 325-a
PEDATE: palmately divided or cleft with the lateral segments again divided.
PEDICEL: the stem of an

individual flower in an inflorescence.
Pedicularis
☐ *canadensis*, 196
PEDUNCLE: the main flower stalk of an inflorescence, supporting a solitary flower or a cluster of flowers. Fig. 5
Peltandra
☐ *virginica*, 32
PELTATE: attached to its stalk near the center of the lower surface instead of at the margin. Fig. 512

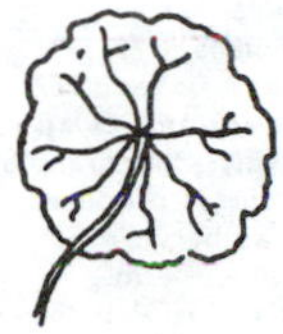

Fig. 512

Pennsylvania Bitter-cress, 116
Penny-cress
  Field, 114
  Perfoliate, 114
Pennyroyal
  Mock, 190
Pennywort, 171
Penstemon, 200
☐ *bradburii*, 201
☐ *cobaea*, 202
  (*grandiflorus*), 201
☐ *hirsutus*, 202
☐ *tubiflorus*, 201
Peppergrass
  Wild, 115
PERFECT FLOWER: a flower with both stamens and pistils.
PERFOLIATE: with the stem appearing to pass through the leaf, Fig. 513-a, or through the fused bases of a pair of leaves, Fig. 513-b (connate-perfoliate)

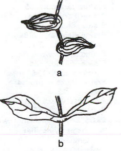

a

b

Fig. 513

Perfoliate Bellwort, 51
Perfoliate Penny-cress, 114
PERIANTH: the collective term for the calyx and corolla together, or for either one if the other is absent.
Periwinkle, 172
PETAL: one member of the inner whorl of non-reproductive flower parts, a division of the corolla. Fig. 5
PETALOID: petal-like, of bracts, sepals or stamens.
Petasites
☐ *frigidus*, 216, 224
  var. *palmatus*, 224
☐ *hybridus*, 224
☐ *sagittatus*, 224
PETIOLATE: with a leaf stalk (petiole).
PETIOLE: the stalk of a leaf. Fig. 2
Phacelia
  Fringed, 184
  Loose-flowered, 184
  Pursh's, 184
  Small-flowered, 184
Phacelia, 182
☐ *bipinnatifida*, 184
☐ *dubia*, 184
☐ *fimbriata*, 184
☐ *purshii*, 184
Philadelphia Fleabane, 225
(*Philodendron*), 33
Phlox
  Chickweed, 180
  Cleft, 180
  Creeping, 179
  Douglas', 179
  Downy, 181
  Hairy, 181
  Hood's, 179
  Moss, 179
  Mountain, 178
  Prairie, 181
  Smooth, 180
  Wild Blue, 179
Phlox
☐ *amoena*, 181
☐ *bifida*, 180
  ssp. *stellaria*, 180
☐ *bryoides*, 179
☐ *divaricata*, 179
☐ *douglasii*, 179
☐ *glaberrima*, 180
☐ *hoodii*, 179
☐ *maculata*, 178
☐ *ovata*, 178
☐ *pilosa*, 181
☐ *stolonifera*, 179
☐ *subulata*, 177
Physalis
☐ *heterophylla*, 194
☐ *virginiana*, 194
PICKEREL-WEED FAMILY, 36
Pimpernel, 169
  Scarlet, 169
  Yellow, 164
(Pineapple), 36
PINEAPPLE FAMILY, 36
Pink
  Carolina, 170
  Fire, 77
  Ground, 177
  Indian, 77, 170

Marsh, 171
Moss, 177
Wild, 78
PINK FAMILY, 77
Pink Lady's-slipper, 60
Pink-root, 170
PINNATE: with veins, lobes or leaflets arranged lengthwise along opposite sides of a leaf or rachis; in a feather-like arrangement. Fig. 3, Fig. 514

Fig. 514

PINNATIFID: pinnately divided.
PISTIL: the female, ovule bearing part of the flower, consisting of ovary, style and stigma. Fig. 5, Fig. 515

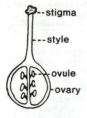

--stigma

---style

-ovule

-ovary

Fig. 515

PISTILLATE: bearing female reproductive parts (pistils) only. Fig. 9
Pitcher Plant, 122
PITCHER-PLANT FAMILY, 122
PITH: the central spongy tissue of a stem.
PLANTAGINACEAE, 203
Plantago
☐ *aristata*, 203
☐ *cordata*, 205
☐ *lanceolata*, 204
☐ *major*, 205
☐ *patagonica*, 204
(*purshii*), 204
☐ *rugelii*, 205
Plantain
Common, 215
English, 204
Heart-leaved, 205
Large-bracted, 203
Robin's, 224
Rugel's, 205
Water, 205
PLANTAIN FAMILY, 203
Plantain-leaved Pussy-toes, 216

plant groups, 6
Plum
Ground, 135
Moxie, 166
PLUMOSE: feathery, like a plume.
Podophyllum
☐ *peltatum*, 103
Pogonia
Rose, 65
Whorled, 64
Pogonia
☐ *ophioglossoides*, 65
(Poinsettia), 145
Pointed Blue-eyed Grass, 56
(Poison Hemlock), 162
poisonous plants, 8-10
POLEMONIACEAE, 177
Polemonium
☐ *reptans*, 177
☐ *van-bruntiae*, 177
POLEMONIUM FAMILY, 177
POLLEN: the granular structures produced in the anthers which contain the sperm.
POLLINATION: the transfer of pollen from the anther to the stigma.
pollinators, 5
Polygala
☐ *paucifolia*, 144
☐ *polygama*, 145
☐ *senega*, 145
POLYGALACEAE, 144
POLYGONACEAE, 72
Polygonatum
☐ *biflorum*, 38
☐ *pubescens*, 39
Polygonum
☐ *aviculare*, 73
☐ *convolvulus*, 72
☐ *penslyvanicum*, 73
☐ *persicaria*, 72
Polytaenia
☐ *nuttallii*, 164
Pomme Blanche, 136
PONTEDERIACEAE, 36
Poor-man's-pepper, 115
POPPY FAMILY, 103
PORTULACACEAE, 76
Potato Dandelion, 215
Potentilla, 127
☐ *anserina*, 128
☐ *argentea*, 129
☐ *canadensis*, 128
☐ *recta*, 129
Prairie False Dandelion, 214
Prairie Iris, 55
Prairie Larkspur, 86
Prairie Parsley, 164
Prairie Phlox, 181
Prairie Ragwort, 218
Prairie-smoke, 131
Prairie Spiderwort, 34
Prairie Turnip, 136
Prairie Violet, 154
Prairie Wake-robin, 44
pressing, 12
PRIMROSE FAMILY, 167
PRIMULACEAE, 167
PROSTRATE: growing with stems lying flat on the ground.
Prostrate Blue Violet, 156
Prostrate Vervain, 189
Prunella
☐ *vulgaris*, 191

Psoralea
☐ *esculenta*, 136
(psyllium seed), 204
PUBERULENT: covered with many short, soft, fine hairs, as in the peach.
PUBESCENCE: the collective term for hairs.
PUBESCENT: with hairs; sometimes used specifically to indicate soft hairs.
Puccoon
Hairy, 188
Hoary, 188
Narrow-leaved, 187
Pulsatilla
☐ *patens*, 97
Purple Avens, 131
Purple Cress, 120
Purple Meadow Rue, 96
Purple-rocket, 117
Pursh's Phacelia, 184
Purslane Family, 76
Purslane Speedwell, 199
Pussy-toes, 216
Plantain-leaved, 216
Puttyroot, 69
Pyrrhopappus
☐ *carolinianus*, 213
☐ *grandiflorus*, 213

Q

Quaker-ladies, 206
Quickweed, 221

R

Rabbit-foot Clover, 139
RACEME: an inflorescence with a long central stalk bearing flowers on smaller stalks along its length. Fig. 10, Fig. 516

Fig. 516

Racemed Milkwort, 145
RACHIS: the central stalk of a pinnately compound leaf or of an inflorescence.
RADIALLY SYMMETRICAL: with parts arranged around a central axis in such a way that the flower can be divided into equal parts along several different diameters; actinomorphic

Fig. 471, Fig. 498
Ragwort
Golden, 218
Prairie, 218
Woolly, 218
Rain Lily, 54
Ramps, 41
Ram's Head Lady's-slipper, 61
RANUNCULACEAE, 85
Ranunculus, 88
☐ *abortivus*, 94
☐ *acris*, 91
☐ *bulbosus*, 90
☐ *fascicularis*, 92
☐ *flabellaris*, 89
☐ *gmelinii*, 89
☐ *hispidus*, 93
☐ *recurvatus*, 90
☐ *repens*, 91
☐ *rhomboideus*, 93
☐ *sceleratus*, 89
☐ *septentrionalis*, 92
RAY: one of the flower stalks in an umbel. Fig. 533
RAY FLOWER: the irregular, tongue-shaped flower of the Compositae. Fig. 443, Fig. 444
RECEPTACLE: the expanded tip of the peduncle or pedicel which bears the floral parts. Fig. 5, Fig. 443
RECURVED: curved backwards. Fig. 517

Fig. 517

Red Baneberry, 94
Red Seeded Dandelion, 213
reference books, v, 9
REFLEXED: bent backwards. Fig. 518

Fig. 518

REGULAR: with all parts uni-
form in size, form, and
union; actinomorphic;
radially symmetrical.
Fig. 519

Fig. 519

RENIFORM: kidney-shaped.
reproduction, 5
REVOLUTE: rolled back-
wards or under with the
tip or margin inside the
roll. Fig. 520

Fig. 520

(Rheum rhaponticum), 74
RHIZOME: an underground
stem, usually horizontal.
(Rhubarb), 74
Robin-running-in-the-hedge,
192
Robin's Plantain, 224
Rock-cress
Hairy, 119

Lyre-leaved, 118
Smooth, 119
Toothed, 119
ROCKROSE FAMILY, 148
root, 1
ROOTSTOCK: an under-
ground stem; rhizome.
ROSACEAE, 126
ROSE FAMILY, 126
Rose Pogonio, 65
ROSETTE: a circle of leaves,
usually basal, as in the
dandelion.
Rose Verbena, 189
Rosy Trillium, 47
ROTATE: with a short tube
and perpendicularly
spreading lobes. Fig.
521

Fig. 521

Rough False Dandelion, 213
Rough-fruited Cinquefoil,
129
Rough-fruited Disporum, 52
Round-leaf Squaw-weed, 219
Round-leaved Orchis, 67
Round-leaved Yellow Violet,
150
RUBIACEAE, 205
Rue
Goat's, 134
Meadow
Early, 95
Purple, 96
Waxy, 96
Rue Anemone, 96
Rugel's Plantain, 205
Rumex
☐ acetosella, 73
☐ altissimus, 74
☐ crispus, 74
☐ obtusifolius, 74
(patientia), 74
RUNNER: a horizontal stem
which produces young
plants at the nodes or
tip; a stolon or rhizome.
(Rutabaga), 111

S

(Sabatia), 171
(Sacred Lotus), 85
Sage, (191)
Lyre-leaved, 191
Scarlet, 191
Sagittaria
Narrow-leaved, 30
Sagittaria
☐ graminea, 30
☐ latifolia, 30
var. pubescens, 30
SAGITTATE: shaped like an
arrow head; triangular
with backwards point-

ing lobes. Fig. 1, Fig.
522

Fig. 522

SALVERFORM: with a long
tube and the lobes
spread perpendicularly.
Fig. 432-a, Fig. 519
Salvia
☐ lyrata, 191
(officinalis), 191
☐ splendens, 191
Sand Violet, 154
SANDALWOOD FAMILY, 71
Sandwort
Blunt-leaved, 81
Thyme-leaved, 80
Sanguinaria
☐ canadensis, 102
Sanicle, 161
Sanicula
☐ canadensis, 161
☐ marilandica, 161
SANTALACEAE, 71
SAPROPHYTE: a plant which
lives on decaying organ-
ic matter.
Sarracenia
☐ flava, 122
☐ purpurea, 122
SARRACENIACEAE, 122
Sarsaparilla, (160)
Wild, 160
Sauce-alone, 117
Saxifraga
☐ pensylvanica, 125
☐ virginiensis, 126
SAXIFRAGACEAE, 123
Saxifrage
Early, 126
Golden, 123
Swamp, 125
SAXIFRAGE FAMILY, 123
SCALE: any of several kinds
of dry or small bracts
or flattened hairs.
SCAPE: a leafless flowering
stalk (peduncle) which
arises directly from the
ground.
Scarlet-fruited Horse Gen-
tian, 210
Scarlet Painted-cup, 195
Scarlet Pimpernel, 169
Scarlet Sage, 191
Scilla
☐ sibirica, 43
Scoot-berry, 53
Scorpion-grass
Spring, 187
Scrophularia
☐ lanceolata, 200
☐ marilandica, 200
SCROPHULARIACEAE, 195

Scutellaria
☐ parvula, 190
☐ serrata, 190
Sedum
☐ ternatum, 123
seed, 5
SEGMENT: one of the lobes
or divisions of a leaf,
leaflet, petal or sepal.
Sego Lily, 49
Self-heal, 191
Seneca Snakeroot, 145
Senecio
☐ aureus, 218
☐ obovatus, 219
☐ pauperculus, 219
☐ plattensis, 218
☐ tomentosus, 218
☐ vulgaris, 216
SEPAL: one of the outermost
parts of the flower,
usually green, covering
the rest of the flower
in bud; a segment of
the calyx. Fig. 5
SEPTUM: a partition, as between halves of fruits
in the Cabbage family.
SERRATE: with forward-
pointing sharp teeth
like those of a saw. Fig.
1
SESSILE: without a stalk.
Sessile-leaved Twisted-stalk,
53
Sheep Sorrel, 73
Shepherd's-purse, 114
Shooting-star, 167
Short-fruited Whitlow-grass,
112
Short-stalked Chickweed, 83
Showy Lady's-slipper, 61
Showy Orchis, 67
Showy Skullcap, 190
Sicklepod, 118
Silene
☐ antirrhina, 80
☐ caroliniana, 78
ssp. pensylvanica, 78
ssp. wherryi, 78
(cucubalus), 78
☐ noctiflora, 79
☐ pratensis, 79
☐ virginica, 77
☐ vulgaris, 78
SILICLE: the short silique-
like fruit of the Cabbage
(Cruciferae) family,
usually with the length
less than twice the
width. Fig. 11
SILIQUE: the long, slender
fruit of the Cabbage
family, in which the two
halves separate from a
median partition. Fig.
11
Silverweed, 128
Silvery Cinquefoil, 129
SIMPLE: composed of one
piece; not compound.
Fig. 2, Fig. 523

Fig. 523

Sinapis
- *alba*, 111
- *arvensis*, 111

SINUATE: a margin which curves inwards and outwards in regular shallow indentations. Fig. 1

Sisymbrium
- *officinale*, 110

Sisyrinchium
- *albidum*, 56
- *angustifolium*, 56
- *atlanticum*, 56
- *campestre*, 56

Skullcap
  Showy, 190
  Small, 190
Skunk Cabbage, 31
Sleepy Catchfly, 80
Slender Blue Flag, 58
Slender Fumewort, 107
Slender-leaved Hedyotis, 208
Slender Toothwort, 113
Smaller Cat's-foot, 217
Smaller Forget-me-not, 187
Small-flowered Bitter-cress, 121
Small-flowered Crowfoot, 94
Small-flowered Nemophila, 181
Small-flowered Phacelia, 184
Small Skullcap, 190
Smartweed, 73
Smilacina
- *racemosa*, 50
- *stellata*, 50
- *trifolia*, 50
(*Smilax*), 160
Smooth Phlox, 180
Smooth Rock-cress, 119
Smooth Yellow Violet, 155
Snakeroot
  Black, 161
  Seneca, 145
(Snapdragon), 195
Snowberry
  Creeping, 166
Snow Trillium, 47
Soapweed, 55
SOLANACEAE, 194
Solanum
- *carolinense*, 194
- *dulcamara*, 194
SOLITARY: occurring singly.
Solomon's-seal
  False, 50
  Great, 38
  Hairy, 39
  Star-flowered, 50
  Three-leaved, 50
Sorrel
  Ladies', 141
  Sheep, 73
  Wood
    Creeping, 141
    Upright Yellow, 141
    Violet, 141

Southern Blue Flag, 59
Southwestern Bedstraw, 209
Spade-leaf Violet, 154
SPADIX: the fleshy spike-like, elongate or globose inflorescence in the Arum family. Fig. 18, Fig. 75-a
Spanish Bayonet, 55
Spanish Moss, 36
SPATHE: an enlarged bract accompanying, and often arched over, a spadix or other inflorescence. Fig. 18, Fig. 75-b
Spatterdock, 84
SPATULATE: somewhat spoon-shaped, with a broadened, rounded apical portion and a tapering base. Fig. 524

Fig. 524

Spear-leaved Yellow Violet, 156
species, 6, 7
specimen, 10-11, Fig. 12
specimen label, 11, Fig. 13
Speckled Wood-lily, 42
(*Specularia perfoliata*), 211
Speedwell
  Common, 197
  Corn, 199
  Purslane, 199
  Thyme-leaved, 198
  Water, 197
SPICATE: spike-like.
Spiderwort
  Bracted, 34
  Ohio, 35
  Prairie, 34
  Virginia, 35
SPIDERWORT FAMILY, 34
Spigelia
- *marilandica*, 170
SPIKE: an elongate inflorescence with sessile flowers arranged along the stem. Fig. 10, Fig. 525

Fig. 525

SPIKELET: a small spike-like inflorescence with in-

conspicuous flowers and many bracts, characteristic e.g. of the Grass family. Fig. 19
Spiranthes
- *lacera*, 68
- *vernalis*, 68
Spreading Chervil, 162
Spring Avens, 130
Spring Beauty
  Carolina, 76
  Virginia, 76
  Western, 76
Spring Coral-root, 63
Spring Cress, 120
Spring Scorpion-grass, 187
SPUR: a hollow, tubular, backward extension of some part of the flower, as in the lower petals of violets.
Spurge
  Alleghany Mountain, 147
  Cypress, 146
  Flowering, 145
  Tinted, 146
SPURGE FAMILY, 145
Squaw-root, 203
Squaw-weed
  Round-leaf, 219
Squill, 43
Squirrel Corn, 106
ssp.: subspecies.
STAMEN: a male, pollen producing part of the flower, consisting of anther and filament. Fig. 5, Fig. 474
STAMINATE: bearing male parts (stamens) only.
STAMINODE: a modified, non-functional, stamen.
Starch Grape Hyacinth, 39
Starflower, 168
Star-flowered Solomon's-seal, 50
Star-grass
  Yellow, 54
Star-of-Bethlehem
  Drooping, 43
Stellaria, 81
- (*aquatica*), 82
- *longifolia*, 82
- *longipes*, 82
- *media*, 81
STELLATE: star-shaped; of hairs with a central stalk with branches radiating from the tip. Fig. 526

Fig. 526

stem, 1
Stemless Lady's-slipper, 60
Stemless Trillium, 45

Sticky Mouse-ear Chickweed, 83
STIGMA: the part of the pistil that receives the pollen. Fig. 7, Fig. 527

Fig. 527

STIPULAR LINE: a scar left in the stem by deciduous stipules.
STIPULAR SHEATH: a thin envelope around the stem formed by the stipules. Fig. 528

Fig. 528

STIPULE: one of a pair of appendages at the base of a petiole, leaf-like or much modified. Fig. 2, Fig. 529

Fig. 529

STOLON: a horizontal above-ground stem which roots at the nodes.
Stonecrop
  Wild, 123
STONECROP FAMILY, 123
Stork's-bill, 142
Strawberry
  Barren, 127
  Dry, 127
  Indian, 127
  Wild Virginia, 126
Streptopus
- *amplexifolius*, 53
- *roseus*, 53
Striped Coral-root, 64

Striped Garlic, 41
STYLE: the part of the pistil that connects the ovary and stigma. Fig. 7, Fig. 530

Fig. 530

Stylophorum
☐ *diphyllum*, 104
SUBTEND: to arise just beneath another structure.
SUBULATE: narrow and tapering upward from the base to a slender point. Fig. 1, Fig. 531

Fig. 531

(Sundew), 122
SUPERIOR OVARY: an ovary that is free from the perianth and positioned above the point of perianth attachment. Fig. 532.

Fig. 532

Swamp Buttercup, 92
Swamp Saxifrage, 125
Sweet Cicely
    Woolly, 162
Sweet Coltsfoot, 216
Sweet Flag, 31
Sweetpea
    Wild, 134
Sweet-scented Bedstraw, 209
Sweet White Violet, 151
Sweet William
    Wild, 178

T

Taenidia
☐ *integerrima*, 164
Tall Buttercup, 91
TAP ROOT: a single enlarged root like a carrot.
Taraxacum
    (*erythrospermum*), 213
☐ *laevigatum*, 213
☐ *officinale*, 213
TEETH: small projections from a margin.
TENDRIL: a twisting, thread-like appendage, adapted for clinging.
Tennessee Milk-vetch, 135
TEPAL: either a sepal or petal in flowers in which the sepals and petals are similar, as in the lily.
Tephrosia
    (*cinerea*), 134
☐ *virginiana*, 134
TERMINAL: at the end of a stem or branch.
TERNATE: in threes.
Texas Bluebonnet, 136
Thalictrum
☐ *dasycarpum*, 96
☐ *dioicum*, 95
☐ *revolutum*, 96
☐ *thalictroides*, 96, 100
Thaspium
☐ *barbinode*, 165
Thimbleweed, 99
Thlaspi
☐ *arvense*, 114
☐ *perfoliatum*, 114
threatened plants, 14-16
Three-leaved Solomon's-seal, 50
Three-lobe Evening Primrose, 159
Thyme-leaved Bluets, 206
Thyme-leaved Sandwort, 80
Thyme-leaved Speedwell, 198
Tiarella
☐ *cordifolia*, 125
Tillandsia
☐ *usneoides*, 36
Tinker's Weed, 210
Tinted Spurge, 146
Toadflax
    Bastard, 71
    Blue, 195
Toadshade, 45
TOMENTOSE: densely woolly, with wavy pressed down hairs.
TOOTHED: with teeth.
Toothed Rock-cress, 119
Toothleaf Evening Primrose, 158
Toothwort
    Cut-leaved, 113
    Large, 113
    Slender, 113
    Two-leaved, 113
Tower Mustard, 119
Tradescantia
☐ *bracteata*, 34

☐ *fluminensis*, 35
☐ *occidentalis*, 34
☐ *ohiensis*, 35
☐ *virginiana*, 35
Trailing Arbutus, 167
Treacle Mustard, 109
Trefoil
    Marsh, 171
Trientalis
☐ *borealis*, 168
    ssp. *latifolia*, 168
TRIFOLIOLATE: with three leaflets.
Trifolium, 138
☐ *arvense*, 139
☐ *aureum*, 139
☐ *campestre*, 139
☐ *hybridum*, 140
☐ *incarnatum*, 139
☐ *pratense*, 140
    (*procumbens*), 139
☐ *repens*, 140
Trillium
    Bashful, 47
    Dwarf White, 47
    Large-flowering, 46
    Nodding, 46
    Painted, 48
    Rosy, 47
    Snow, 47
    Stemless, 45
Trillium
☐ *catesbaei*, 47
☐ *cernuum*, 46
☐ *erectum*, 45
☐ *grandiflorum*, 46
☐ *nivale*, 47
☐ *recurvatum*, 44
☐ *sessile*, 45
☐ *undulatum*, 48
Triodanis
☐ *lepto·arpa*, 211
☐ *perfoliata*, 211
Triosteum
☐ *aurantiacum*, 210
☐ *perfoliatum*, 210
Trollius
☐ *laxus*, 101
Trout Lily, 41
True Watercress, 116
Trumpets, 122
TRUNCATE: blunt at the base or apex as if cut off. Fig. 1
TUBER: a thickened underground storage stem.
TUBEROUS: with tubers or tuber-like.
Tufted Loosestrife, 170
Turnip, (111)
    Indian, 33
    Prairie, 136
Tussilago
☐ *farfara*, 217
Twayblade
    Heart-leaved, 69
    Lily-leaved, 70
    Loesel's, 70
TWINING: descriptive of a vining plant which climbs by twisting around a support.
Twinleaf, 102
Twisted-stalk
    Clasping-leaved, 53
    Sessile-leaved, 53
Two-leaved Toothwort, 113

U

UMBEL: an inflorescence with all of the pedicels arising from the same point. Fig. 10, Fig. 337, Fig. 533

Fig. 533

UMBELLIFERAE, 160
UNISEXUAL: bearing reproductive parts of only one sex.
UNITED: grown together.
Upright Yellow Wood Sorrel, 141
Uvularia
☐ *grandiflora*, 51, 52
☐ *perfoliata*, 51, 52
☐ *puberula*, 51
☐ *sessilifolia*, 51

V

Valerian
    Greek, 177
var., variety, 6
VARIEGATED: containing several colors.
VASCULAR BUNDLE: a strand of conducting tissue.
VEIN: one of the strands of conducting tissue in a leaf or flower part.
Veiny Pea, 134
(Venus' Flytrap), 122
Venus's Looking-glass, 211
Veratrum
    (*californicum*), 50
☐ *viride*, 31, 49
Verbena
    Garden, 189
    Rose, 189
Verbena
☐ *bracteata*, 189
☐ *canadensis*, 189
☐ X *hybrida*, 189
VERBENACEAE, 189
Vernal Whitlow-grass, 112
Veronica, 196
☐ *americana*, 197
☐ *anagallis-aquatica*, 197
☐ *arvensis*, 199
☐ *officinalis*, 197
☐ *peregrina*, 199
☐ *persica*, 198
☐ *serpyllifolia*, 198
Vervain
    Prostrate, 189
VERVAIN FAMILY, 189
VESICLE: a small fluid-filled

(or gas-filled) sac.
Vesicular Corydalis, 108
Vetch, 132
  Carolina, 132
  Cow, 133
  Hairy, 132
  Milk-
    Low, 135
    Tennessee, 135
Vicia
☐ *americana*, 132
☐ *caroliniana*, 132
☐ *cracca*, 133
☐ *villosa*, 132
VILLOUS: densely hairy with
  long soft hairs.
Vinca
☐ *minor*, 172
Viola, 149
☐ *adunca*, 157
☐ *affinis*, 152
☐ *arvensis*, 155
☐ *(bicolor)*, 155
☐ *blanda*, 151
☐ *canadensis*, 157
☐ *conspersa*, 156
☐ *(cucullata)*, 153
☐ *hastata*, 156
☐ *incognita*, 151
☐ *(kitaibeliana)*, 155
☐ *lanceolata*, 150
☐   ssp. *vittata*, 150
  macloskeyi
☐   ssp. *pallens*, 151
☐ *missouriensis*, 152
☐ *obliqua*, 153
☐ *(pallens)*, 151
☐ *palmata*, 154
☐ *pedata*, 151
☐ *pedatifida*, 154
☐ *(pensylvanica)*, 155
  pubescens, 155
☐   var. *eriocarpa*, 155
☐   var. *pubescens*, 155
☐ *rafinesquii*, 155
☐ *rostrata*, 156
☐ *rotundifolia*, 150
☐ *sagittata*, 154
☐ *sororia*, 153
☐ *striata*, 157
☐ *tricolor*, 155
☐ *tripartita*, 156
☐ *walteri*, 156
☐ X *wittrockiana*, 155
VIOLACEAE, 149
Violet
  Affinis, 152
  American Dog, 156
  Arrow-leaved, 154
  Birdfoot, 151
  Blue Marsh, 153
  Bog, 153
  Canada, 157
  Common Blue, 153
  Confederate, 153
  Downy Yellow, 155
  Early Blue, 154
  Green, 149
  Lance-leaved, 150
  Long-spurred, 156

  Missouri, 152
  Pale, 157
  Prairie, 154
  Prostrate Blue, 156
  Round-leaved Yellow, 150
  Sand, 154
  Smooth Yellow, 155
  Spade-leaf, 154
  Spear-leaved Yellow, 156
  Sweet White, 151
  Water, 150
  Wild White, 151
  Woolly Blue, 153
  Yellow, 155
Violet
  Dame's, 116
Violet
  Dogtooth, 41
VIOLET FAMILY, 149
Violet Wood Sorrel, 141
Virginia Cowslip, 185
Virginia Goatsbeard, 215
Virginia Ground Cherry, 194
Virginia Spiderwort, 35
Virginia Spring Beauty, 76
Virginia Waterleaf, 183
VISCID: sticky.

W

Waldsteinia
☐ *fragarioides*, 127
Wake-robin
  Ill-scented, 45
  Prairie, 44
Wandering Jew, 35
Wapato, 30
Water-carpet, 123
Watercress
  Native, 116
  True, 116
Water-crowfoot
  Yellow, 89
Water Hyacinth, 36
Waterleaf
  Appendaged, 182
  Canada, 183
  Large-leaved, 183
  Virginia, 183
WATERLEAF FAMILY, 181
Waterlily
  Fragrant, 85
WATERLILY FAMILY, 84
Water Mouse-ear Chickweed,
  82
Water Plantain, 205
WATER-PLANTAIN FAMILY, 30
Water Speedwell, 197
Water Violet, 150
Waxy Meadow Rue, 96
Wedge-leaved Whitlow-
  grass, 112
Western Daisy, 222
Western Dwarf Dandelion,
  214
Western Spring Beauty, 76
Western Wallflower, 109
Wild Bleeding-heart, 105
Wild Blue Phlox, 179

Wild Buckwheat, 72
Wild Calla, 32
Wild Columbine, 87
Wild Comfrey, 185
Wild Four-o'clock, 75
Wild Geranium, 143
Wild Ginger, 71
Wild Hyacinth, 43
Wild Indigo
  Atlantic, 138
  Large-bracted, 137
Wild Leek, 41
Wild Lemon, 103
Wild Lily-of-the-valley, 38
Wild Liquorice, 209
Wild Lupine, 136
Wild Mandrake, 103
Wild Oats, 51
(Wild Onion), 48
Wild Pansy, 155
Wild Peppergrass, 115
Wild Pink, 78
Wild Sarsaparilla, 160
Wild Stonecrop, 123
Wild Sweetpea, 134
Wild Sweet William, 178
Wild Virginia Strawberry,
  126
Wild White Violet, 151
Whippoorwill-shoe, 62
White Avens, 130
White Baneberry, 95
White Campion, 79
White Clintonia, 42
White Clover, 140
White Evening Primrose, 159
White Fawn-lily, 41
White Lady's-slipper, 62
White Mustard, 111
White Sweet Clover, 137
Whitlow-grass
  Carolina, 112
  Short-fruited, 112
  Vernal, 112
  Wedge-leaved, 112
WHORL: a ring of parts all
  arising at the same level
  on an axis, especially 3
  or more leaves in a
  circle at the node. Fig.
  4, Fig. 534

Fig. 534

Whorled Milkweed, 174
Whorled Pogonia, 64

Windflower, 98, 99
WING: a ridge or thin bor-
  der projecting from an
  angle or margin of an
  organ.
Wintergreen, 166
  Flowering, 144
Wister's Coral-root, 63
Wood Anemone, 99
Wood Betony, 196
Wood-lily
  Speckled, 42
WOOD SORREL FAMILY, 141
(Woolly Blue Violet), 153
Woolly Indian Wheat, 204
Woolly Ragwort, 218
Woolly Sweet Cicely, 162
Worm-grass, 170
Wormseed Mustard, 109

Y

Yarrow, 222
Yellow Adder's-tongue, 40
Yellow Bedstraw, 208
Yellow Clintonia, 42
Yellow Clover, 139
Yellow Cress, 110
Yellow Lady's-slipper, 62
Yellow Mandarin, 52
Yellow Pimpernel, 164
Yellow-rocket, 110
Yellow Star-grass, 54
Yellow Violet, 155
Yellow Water-crowfoot, 89
Yucca
☐ *filamentosa*, 55
☐ *glauca*, 55

Z

Zephyranthes
☐ *atamasco*, 54
Zephyr Lily, 54
Zigadenus
☐ *nuttallii*, 48
Zizia
☐ *aptera*, 163
☐ *aurea*, 165
ZYGOMORPHIC: bilaterally
  symmetrical; irregular;
  capable of being di-
  vided into equal halves
  along only one plane.
  Fig. 535

Fig. 535

NOTES

NOTES